High and Dry

High and Dry

Meeting the Challenges of the World's Growing Dependence on Groundwater

William M. Alley and Rosemarie Alley

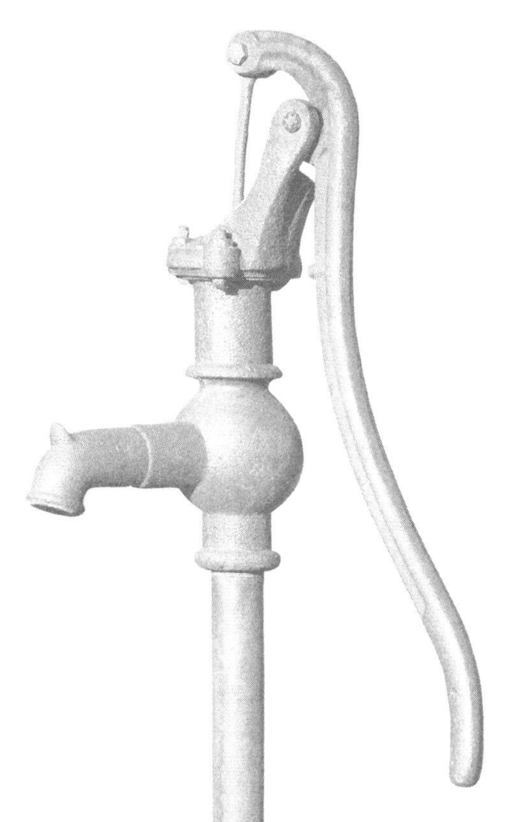

Yale UNIVERSITY PRESS/NEW HAVEN & LONDON

Published with assistance from the foundation established in
memory of Philip Hamilton McMillan of the Class of 1894, Yale College.

Yale University Press books may be purchased in quantity for educational,
business, or promotional use. For information, please e-mail
sales.press@yale.edu (U.S. office) or sales@yaleup.co.uk (U.K. office).

Designed by Mary Valencia.

Set in Minion type by Westchester Publishing Services.
Printed in the United States of America.

ISBN 978-0-300-22038-4 (hardcover : alk. paper)

Library of Congress Control Number: 2016949267

A catalogue record for this book is available from the British Library.

This paper meets the requirements of ANSI/NISO Z39.48–1992
(Permanence of Paper).

10 9 8 7 6 5 4 3 2 1

To our grandchildren, Oliver and Julia.
May there be plentiful and clean groundwater for them,
their children, and their grandchildren.

Contents

Acknowledgments

This book would never have been written without the help and insights of many individuals who care about groundwater and its role in people's lives and the environment. Joseph Calamia at Yale University Press made many invaluable edits and suggestions and guided us through each step in the publication process. Samantha Ostrowski helped with the figures and manuscript preparation. Julie Carlson gave the book a thorough copyediting. Mike Campana and Mark Giordano read a draft and provided many useful comments and suggestions. We are grateful for their insightful reviews. Sharon Megdal patiently taught us about the complexities of Arizona water. Kathleen Ferris, Paul Polak, Holly Richter, and Marvin Glotfelty all generously gave their time for interviews. Joe Ayotte, Mark Borchardt, Tom Cech, Devin Galloway, Rochelle Holm, Randy Hunt, Dick Jackson, Laurel Lacher, Stan Leake, Jim Leenhouts, Alan MacDonald, Rebecca Nelson, Laurie Rardin, Michelle Sneed, Paul Susca, and Pietro Teatini commented on various chapters and helped with fact-checking. Any errors that remain are ours alone. Kenley Bairos, Nathaniel Delano, and Sally House provided assistance with key illustrations. Finally we appreciate the contributions of hydrologists with the U.S. Geological Survey and members of the National Ground Water Association, with whom we've had numerous productive discussions.

Beyond Rain

Heaven is under our feet as well as over our heads.
—Henry David Thoreau

In November 2007, Georgia's governor, Sonny Perdue, stood behind the podium at the state capitol and gazed at the crowd. Hundreds of people from all across the state had traveled to be here on this late fall day. The choir ended their hymn. Silence ensued. All eyes were riveted on their governor.

Perdue stepped to the microphone and began to speak. "We've come together here simply for one reason and one reason only: To very reverently and respectfully pray up a storm . . . It's time to appeal to Him who can and will make a difference."[1]

Members of the secular Atlanta Freethought Society gathered at the edge of the crowd, staunchly protesting this public prayer vigil. Ed Buckner, who had organized the protest, explained the society's views to anyone willing to listen: "The governor can pray when he wants to. What he can't do is lead prayers in the name of the people of Georgia."[2]

Can or *can't,* in the minds of most Georgians division of church and state was far from the issue. The southeastern United States was in the grip of one of the worst droughts in decades—possibly in a century. Parts of the southeast hadn't seen normal rainfall for a year and a half. Some areas were forty inches, basically a year's worth of rain, below normal. Rivers had sunk to dangerously low levels. The boat ramp at Lake Lanier reservoir looked like a diving board. The drought was now threatening public water supplies.[3]

The citizens of Georgia already had heeded the governor's plea for strict water conservation. After declaring a state of emergency in much of the state, Perdue had battles to fight. First he filed suit against the U.S. Army Corps of Engineers, the federally appointed agency that basically calls the shots on reservoirs in the eastern United States. The governor wanted the Army Corps to reduce the amount of water being released from Georgia's Lake Lanier for Alabama and Florida (and downstream endangered species). The Corps flatly refused. He appealed to President George W. Bush to put some pressure on the Corps, but the president wasn't of a mind to anger all those voters downstream in Florida. Finally, Perdue met with the respective governors to try to work out some kind of deal. But the only thing they accomplished was a further escalation of the tri-state water war that had been alternately simmering and boiling for nearly two decades. At this point, it seemed there was only one thing left to try. And so the governor prayed.

Two weeks earlier, a gospel concert dedicated to rain had attracted hundreds at an Atlanta church. Days later, a prayer rally at a high school football stadium in Watkinsville, Georgia, drew another large crowd. Nor was Perdue the first governor to try prayer. In July, Alabama's governor, Bob Riley, had declared an entire week "Days of Prayer for Rain." It didn't work. By now, worry had turned to fear.

Battles over interstate water are a business-as-usual part of life in the arid western United States, but Easterners have tended to take their water wealth for granted. In the past few decades, this has begun to change. The fastest-growing eastern cities need additional water supplies and are discovering the difficulties of trying to secure water that's already allocated. One of these cities is the Belle of the Old South—Atlanta, Georgia.

Since the 1970s, Atlanta's spectacular growth has been the driving force behind the battle officially dubbed the "Water War of the South." The seeds of this conflict go back to the 1950s when the dam that created Lake Lanier was constructed. Lying less than fifty miles (eighty kilometers) north of Atlanta, the lake was to supply much of the city's water—along with allocations for the downstream states of Alabama and Florida. When these allotments were being discussed, Atlanta was still a small, leisurely paced, southern city. Having no way of reading the tea leaves, much of Lake Lanier's water was dedicated to hydropower.

This situation changed in the 1970s, as Atlanta's population began to explode into a large metropolitan area and state officials realized they needed more water. The problem was, that big reservoir (almost twice the size of Manhattan) in their backyard was already spoken for. So they turned to Congress. Under its constitutional authority to regulate interstate commerce, Congress authorized the U.S. Army Corps of Engineers to study ways to supply that water.

In 1989, the Corps announced that it would reallocate some of the water stored for hydropower in Lake Lanier for Atlanta's public supply, basically doubling its allotment. Naturally, Georgians heartily applauded the Corps' decision. It was plain as day that they had a right to the water in their own state. But the people of Alabama and Florida were outraged. They had their own population growth and economic goals to consider. Freshwater flows from Lake Lanier also were essential to Florida's booming oyster industry in Apalachicola Bay. Alabama and Florida initiated a lawsuit against the Corps.

When tempers cooled and the sobering reality of court decisions yielding uncertain outcomes began to sink in, the three states decided to negotiate their own water compact. Or at least try to. The three states agreed to a comprehensive study of the water resources in the Apalachicola-Chattahoochee-Flint River Basin, mercifully shortened to the ACF River Basin.

They got closer to an agreement than many would have predicted. By 1997, the three states had signed into federal law the ACF River Basin Compact, which was basically an agreement to agree on an allocation formula. Regardless of how silly that may sound, this compact was an important first step. Congress approved it. President Clinton signed federal authorizing legislation with a negotiation deadline of December 31, 1998. But these sorts of deadlines have a way of coming and going. In July 2003 (three deadlines later), the governors signed a Memorandum of Understanding that set a blueprint for water allocation—an outline, if you will, yet another vital step along the way. The basic plan was that Atlanta would receive 705 million gallons (2.7 million cubic meters) of water a day from Lake Lanier, with the possibility of even greater withdrawals in the future—while Alabama and Florida would have water security in the form of minimum flow requirements.

Everything appeared to be going smoothly until Florida governor Jeb Bush started having second thoughts. The deal looked good on paper, but who was to say the required minimum flows wouldn't sooner or later become the norm? Other problems soon began to worry one or another of the governors. After more than six years of compact negotiations, and thirteen years after the lawsuit that began it all, the compact breathed its last gasp on August 31, 2003. It wasn't long before the governors were talking to each other only through their lawyers.

Meanwhile, the drought hit, the governor prayed, and rain poured from the sky on believers and nonbelievers alike during Atlanta's annual Thanksgiving Day marathon. When the drought finally loosened its grip, permanent conservation measures were enacted.[4] Many people in this part of the water-rich east still remember the drought, and how so much depends on the highly fickle phenomenon of rain.

Praying for rain is just one of the many traditions that humans have devised in order to continue a settled way of life. Other approaches, involving highly ritualized ceremonials or exacting chemical potions, fall within a general category known as rainmaking. The goal is to get rain, but not too much.

Probably no one has better demonstrated the problem of overdoing it than California's most famous rainmaker, Charles Hatfield. Active in the early 1900s, Hatfield was a character and looked the part with his collarless white dress shirt, jauntily perched fedora, and twinkle in his eye. Paul Newman's role in *Butch Cassidy* could have been modeled after Hatfield.

Hatfield liked to explain that he didn't "create" rain, but merely persuaded nature to release the vast stores of moisture always present in the atmosphere. He also put on one heck of a show. First he would assemble his tall and gangly rainmaking tower. Then he would mix up his "chemical affinity highball," a carefully guarded secret recipe of nearly two dozen chemicals that he aged for several days, then poured into pans on top of the tower where it would evaporate to bring rain. One spectator quipped that the stink was so bad it smelled like a limburger cheese factory and "it rains in self-defense."[5]

Following several modest demonstrations of his talent, Hatfield proposed that for a thousand dollars he would provide the drought-stricken city of Los Angeles with eighteen inches of rain. The offer made newspaper headlines, a deal was struck, and the rains came as promised. Hatfield was rewarded with a spot on the lead float in Pasadena's Rose Bowl Parade. His fame soon spread to Europe. London hailed him as the World's Greatest Rainmaker.[6]

In December 1915, a serious drought was threatening San Diego's water supply. The city council received a recommendation from the Wide Awake Improvement Club to invite Hatfield to San Diego. In what was to become his most famous and potentially most lucrative commission, Hatfield promised to fill one of the city's main reservoirs to overflowing by the next December. If he delivered, the city would pay him $10,000. If he didn't, he wouldn't get paid. It was a simple gentlemen's agreement. Everyone shook hands.[7]

Within a month, record rains had filled the reservoir to overflowing (as promised) and then proceeded to wash out bridges, railroad tracks, streets, houses, and crops. There was loss of life. Damage claims totaled $3.5 million, in 1915 dollars. Someone suggested paying Hatfield $100,000 to quit. Most people wanted to lynch him.

Unruffled, Hatfield demanded his fee. Officials balked; the city attorney wrote the whole thing off to an act of God. Hatfield continued to demand his fee. The city finally agreed to pay him, but only if he accepted responsibility for all the damage claims. The issue languished in court for more than two decades, and was finally dismissed on the grounds that there had been no written contract.

In the late 1880s, another pseudo-scientific approach to rainmaking set the stage for one of the costliest human-induced environmental disasters in world history. Anyone who has made the long drive west across Kansas has noticed the change in topography. Soon after passing Lawrence, the lush green rolling hills give way to a land of endless flatness, with a constant wind buffeting the car.

In the early 1800s, after the Louisiana Purchase, a few intrepid souls went out to take a look at all this new territory. Through the process of unanimous agreement, they named this huge upland the Great American Desert. Zebulon Pike (after whom Pike's Peak is named) crossed this formidable desert all the way to the Rocky Mountains and subsequently wrote in his diary: "Our citizens being so prone to rambling, and extending themselves on the frontiers, will, through necessity, be constrained to limit their extent on the west to the borders of the Missouri and the Mississippi, while they leave the prairies, incapable of cultivation, to the wandering . . . Aborigines of the country."[8] Pike saw a restriction on westward expansion as beneficial to continuation of the union.

Word got out. The mix of bad name, bad press, and actual reality kept most pioneers from venturing into this formidable terrain. The Homestead Act of 1862 attempted to jump-start settlement, but there were few takers.

Then things changed. When the Civil War ended in 1865, the powerful eastern railroad barons asked themselves why they were running trains all the way from coast to coast with nothing to show for it across that whole expanse of flat emptiness. Based on the time-tested premise that you can talk humans into virtually anything, they began an audacious public-relations campaign. First, a name change was in order. The Great American Desert was rechristened the Great Plains. And the Great Plains became nothing less than a land of plenty just waiting to be discovered.

Initial takers numbered in the hundreds, but a stroke of luck changed everything. In the 1870s and early 1880s, parts of Nebraska and Kansas were unusually rainy. Coaxed along by a serious advertising campaign, it wasn't long before homesteaders began to arrive in unprecedented numbers. Hundreds, then thousands, then a virtual stampede of hundreds of thousands of land-hungry farmers were now willing to head west. Within a decade, nearly two million people had settled on the Great Plains.

The rains continued, and numerous theories attempted to explain it. Some suggested that the iron on the railroad or telegraph lines was responsible for the increased rainfall. (The why and how weren't tackled.) Probably the most colorful theory was that normal atmospheric circulation was being disturbed through the concussion of moving trains, thereby causing rain clouds to stop and deliver. There was also the popular theory that forests produce rain—even though there were no forests and tree planting was still in the nursery stage.

Then along came Nebraskan Samuel Aughey, who took a different tack: "It is the great increase in the absorptive power of the soil, wrought by cultivation, which has caused, and continues to cause, an increasing rainfall in the State." There was something about Aughey's theory that, at least, *sounded* like credible science.[9]

It took Charles Dana Wilber, a land speculator and front man for the Burlington Railroad, to boost Aughey's theory to the heights of officialdom. In his 1881 book *The Great Valleys and Prairies of Nebraska and the Northwest,* Wilber wrote that green growing crops, in place of dry, hard-baked earth covered with sparse buffalo grass, would create a cooling condensing surface. "A reduction of temperature must at once occur," he concluded, "accompanied by the usual phenomena of showers. The chief agency in this transformation is agriculture. To be more concise: Rain follows the plow."[10]

In one fell swoop, Dana explained *and* guaranteed a lucrative farming future for Great Plains settlers. There was no quibbling about the soil. The deep, rich loam of this vast prairie was incredible stuff. With rain now officially following the plow, "God speed the plow!" became everyone's slogan. By the early twentieth century, continued waves of immigration had resulted in over 50 million acres (20 million hectares) being turned by the plow. Then God really did speed the plow, as tractors replaced the horses and humans sweating up and down rows of hot dusty fields. Another boon arrived when World War I boosted agricultural prices, leading to a dramatic increase in cultivation. By war's end, nearly 90 million acres had been turned, churned, and planted.

For ten to fifteen years everything was great. Wheat prices were soaring. By 1929, over 100 million acres were under cultivation. It was the best wheat crop ever. When Wall Street collapsed and prices plummeted, the solution to this economic Armageddon was to plow more land and produce more wheat than ever before. In the spring of 1931, the winter wheat stood shoulder high, one of the best crops in twenty-five years, but the going price was half of what it cost to grow it.[11]

Then the unthinkable happened when drought struck with a vengeance. On January 21, 1932, a dust cloud outside Amarillo—the first of many—climbed nearly two miles (three kilometers) into the sky. The dust storms arrived with increasing frequency, and there was no rain between storms. The Dust Bowl's fury lasted nine years. For those who stayed, it became painfully obvious that rain doesn't follow the plow. The Great Plow-Up had plowed up the wrong kind of storm.[12]

By the early 1940s, rain was once again pelting the Great Plains. On the Texas High Plains, 1941 was the wettest year on record. But the Dirty Thirties were not an anomaly. Drought struck again in 1945 and didn't let up for four years. Then came the Filthy Fifties. In spite of soil conservation measures adopted after the Dust Bowl, dust storms were once again blowing over forty million acres (sixteen million hectares) of cultivated soil and putting farmers out of work.[13]

Ironically, the High Plains Aquifer lay right under their feet. Stretching from South Dakota to Texas, this aquifer is one of the largest bodies of groundwater in the world. Yet there was no way to pump enough for any sizable farming operation. Windmills had been ushered in with high hopes in the 1890s, but were severely limited by how much and how deep they could pump.[14]

Everything changed with the widespread adoption of the centrifugal pump. While a windmill could deliver modest amounts of water when the wind was blowing, the centrifugal pump could lift a gushing stream of water 24/7. In places, these high-powered pumps were delivering 1,000 gallons (3,800 liters) or more every minute.[15] This was something like exchanging your Model T for an eight-cylinder Dodge Ram pickup. No longer were farmers dependent on fickle Mother

Nature. No more sleepless nights praying for rain. With the centrifugal pump, they could beat back drought with a readily available and seemingly limitless supply of groundwater. Farmers on the High Plains finally had job security.

Then came center pivot irrigation, which made it possible to irrigate huge swaths of land. Neither of these inventions, however, would have amounted to anything without electricity—compliments of the New Deal rural electrification program begun in the 1930s. In 1919, for example, the number of farms in Nebraska with electric power was about 10 percent. By 1954, approximately 95 percent had electricity.[16] This perfect triad—the centrifugal pump, center pivot irrigation, and electricity—changed farming forever. Even when it rained, irrigated land was producing much greater crop yields than dry land farming ever had. By the 1960s, the Great Plains was emerging as the world's "grain basket" with a $30 billion agricultural and livestock industry—every year, rain or shine. You could count on it.

The tapping of the High Plains Aquifer began a groundwater revolution that would sweep the world. Between 1950 and 1975, rapid expansion in groundwater use occurred in many industrialized countries. In the 1970s, a second wave began in India, Pakistan, China, the Middle East, and North Africa. Currently, a third wave of groundwater development is appearing in sub-Saharan Africa and in Southeast Asian countries such as Cambodia and Vietnam.[17]

Among the most important attributes of groundwater are its wide accessibility and the large volumes of water in storage. Groundwater accounts for about 95 percent of the Earth's unfrozen fresh water and occurs almost everywhere beneath the land surface. This makes groundwater a self-service resource accessible to nearly everyone, instead of just the favored few who live near a river or canal. Groundwater is a democratic resource.

The surge in global groundwater use is largely because of irrigation. Around 70 percent of groundwater pumped is used for agriculture, benefiting the entire continuum from agribusiness to many millions of smallholder, poor farmers.[18]

Groundwater is the primary source of drinking water for about half the world's people, as well as the dominant source of household water in rural areas. Many large cities, including half the world's megacities, depend solely or largely on groundwater. Cities as diverse as Bangkok, Buenos Aires, Cairo, Calcutta, Jakarta, London, Mexico City, and Teheran are all heavily dependent on groundwater for their existence. Groundwater plays a crucial role in the global urbanization phenomena we are now witnessing.[19]

Groundwater is also an integral part of the environment. During dry spells, rivers continue to flow because of groundwater. Many plants and animals depend on groundwater discharge to springs, lakes, wetlands, and estuaries for their survival.

With so much at stake, the importance of groundwater is irrefutable—but the halcyon days of unrestricted use are ending. In many parts of the world, current levels of pumping are an issue of grave concern. Large-scale pumping operations largely started on the High Plains, so it's not surprising that this is where some of the first signs of trouble appeared. Groundwater withdrawals from the High Plains Aquifer exceed those of any other aquifer in the United States. By the 1970s, groundwater levels were steadily declining over large areas. This virtual ocean of groundwater, which accumulated over thousands of years, is being used up in decades.

The High Plains is not a monolith. The northern part of the aquifer has seen little or no depletion. Precipitation is more than adequate to replenish it. In contrast, large swaths of prime farming areas in the central and southern High Plains are in serious trouble. It's no longer a question of *if* these areas will run out of groundwater that is economically feasible to pump, but rather how long it will be until the day of reckoning arrives. If pumping continues at current rates, it is estimated that 35 percent of the southern High Plains will be unable to support irrigation within thirty years.[20] The situation is so dire that some have even suggested letting the region depopulate and return to being a grazing ground for buffalo.

From 1900 to 2008, the groundwater in the United States was depleted by about twice the volume of water contained in Lake Erie.

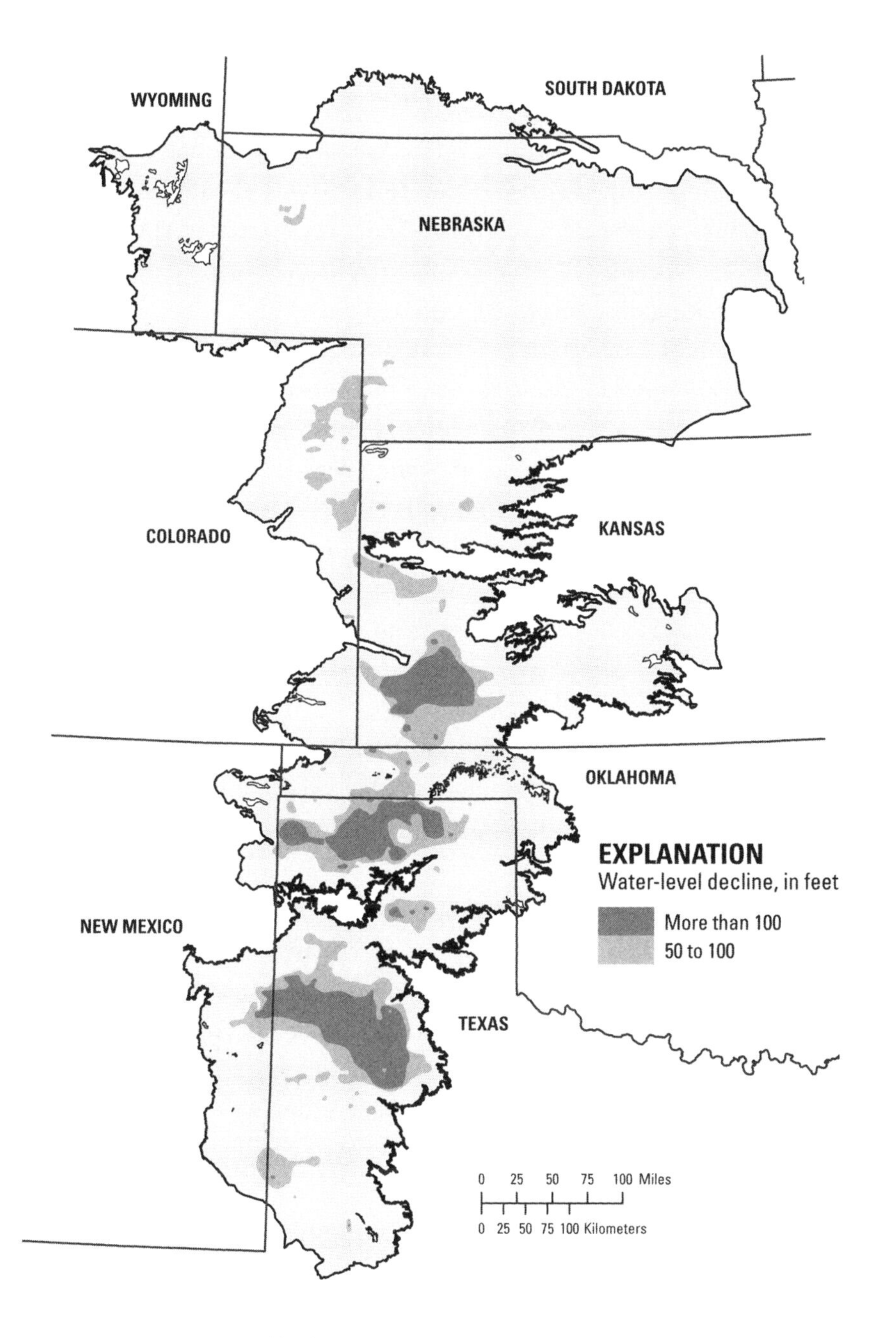

Areas with more than fifty feet of water-level decline in the High Plains Aquifer, predevelopment to 2013. *Source:* V. L. McGuire, U.S. Geological Survey.

Three groundwater systems accounted for most of this depletion: the High Plains Aquifer, the Mississippi Embayment, and California's Central Valley. Most disturbing is that the rate of depletion increased substantially during the period 2000 to 2008, with the fastest rates of depletion taking place in California. And these figures don't take into account the accelerated groundwater depletion during California's record-breaking drought, which began in 2012.[21]

Groundwater depletion is now a global problem. Aquifers providing irrigation water for some of the world's most productive farmland are under stress and face an uncertain future. Meanwhile, global food production may need to double in coming decades to feed nine billion or more people. New crop varieties, improved agricultural practices, reduced food waste, and leaner diets can all help meet this goal, but groundwater is essential. On a global scale, more than 40 percent of consumptive irrigation use comes from groundwater. Moreover, although only 17 percent of croplands are currently irrigated, these lands produce 40 percent of the world's food.[22]

Among the nations with the most significant annual groundwater extractions are those with some of the largest populations and with standards of living dependent, to a great extent, on groundwater. China is a case in point.

The semi-arid North China Plain is one of the most densely populated regions in the world. This rapidly urbanizing area, which includes Beijing and Tianjin, is China's cultural, political, and economic center. It's also a major agricultural region. Leery of depending on imports to feed the country, China's government has long pursued self-sufficiency in grain and other crops. Intensive double-cropping of winter wheat and maize in the North China Plain has been essential for achieving that goal.

The North China Plain has relied on groundwater pumping for more than 70 percent of its water supply. By the end of the 1990s, groundwater levels were declining at a rate of more than three feet (one meter) each year. Groundwater depletion has resulted in severe environmental consequences, including dried-up rivers, shrinking wetlands, land subsidence (sinking or settling of the earth), seawater intrusion, and deterioration of water quality. In 2014, in an effort to reduce pumping, China cut wheat production for the first time in the North China Plain's productive Hebei province. In addition to other efforts to reduce water demand, China has undertaken immense engineering projects to increase the North's water supply.[23]

China's population is split almost evenly between its northern and southern parts, but the country's water distribution is a tale of massive inequality. Eighty percent of the country's water resources are in the southern half. To redistribute this water wealth, China has developed the biggest interbasin transfer scheme in the world—the South-North Water Transfer Project—to bring water from the Yangtze River north to the Yellow River and North China Plain. The price tag for this water-moving megalith exceeds $80 billion. The environmental and social costs from reduced flows in the Yangtze are likewise colossal. The project has displaced hundreds of thousands of people, many of them poor farmers. Two of the water-transport routes are now up and running, each moving water more than seven hundred miles (1,100 kilometers), with a controversial third route in the planning stage.[24]

The project currently has the capacity to deliver 6.6 trillion gallons (twenty-five billion cubic meters) of fresh water per year, much of which goes to China's megacities. Meanwhile, groundwater levels in the North China Plain continue to drop and the country needs to find yet other ways to reduce pumping. The water transfer also does not address major causes of water shortages from inefficient agricultural, industrial, and urban water use. The water-scarcity problems are compounded by rampant groundwater pollution. The poor quality of as much as 60 percent of water in China's northern rivers further reduces the supply of clean water for drinking and domestic use.[25]

China is far from alone. In less than fifty years, pressures from food production and population growth have led to global declines in groundwater supplies that once appeared inexhaustible. Today's practices are also determining groundwater quality for decades to come. David Seckler, former director of the International Water Management

Institute based in Sri Lanka, warned about the consequences for many developing countries: "The penalty of mismanagement of this valuable resource is now coming due, and it is no exaggeration to say that the results could be catastrophic for these countries, and given their importance, for the world as a whole."[26]

To grasp the global extent of this problem, there's nothing like NASA's Gravity Recovery and Climate Experiment, known as GRACE. GRACE relies on the interplay of two satellites (nicknamed Tom and Jerry) that fly in a tandem orbit 137 miles (220 kilometers) apart. The satellites are pulled apart and pushed together as they fly over areas of higher or lower gravity. Each bombards the other with microwaves in order to measure the distance between them to intervals of less than the width of a human hair. Repeat orbits detect changes in gravity that are primarily driven by changes in the volume of water in the Earth below. GRACE is like a giant scale in the sky.[27]

Resulting from collaboration between the United States and Germany, the two satellites were launched in 2002 on the back of a used intercontinental ballistic missile from a Russian space facility. Soon thereafter, this technological marvel began streaming data back to Earth. After adjusting for soil moisture and all forms of surface water, GRACE data can provide precise estimates of changes in groundwater storage, but only over an extremely large area—say, about the size of the state of Nebraska.[28] Consequently, its data can't describe what's happening to water levels in different parts of an aquifer, where the main action is.

GRACE has other limitations. It doesn't indicate where seawater is intruding, or where (and by how much) the land is subsiding. GRACE is silent on how water quality is changing, and how streams and other surface-water bodies are being affected by pumping. Nonetheless, GRACE has done an excellent job of getting people's attention. While most people can't understand a groundwater model or a technical paper, GRACE's user-friendly maps show at a glance where groundwater is being depleted around the world. GRACE has enabled people,

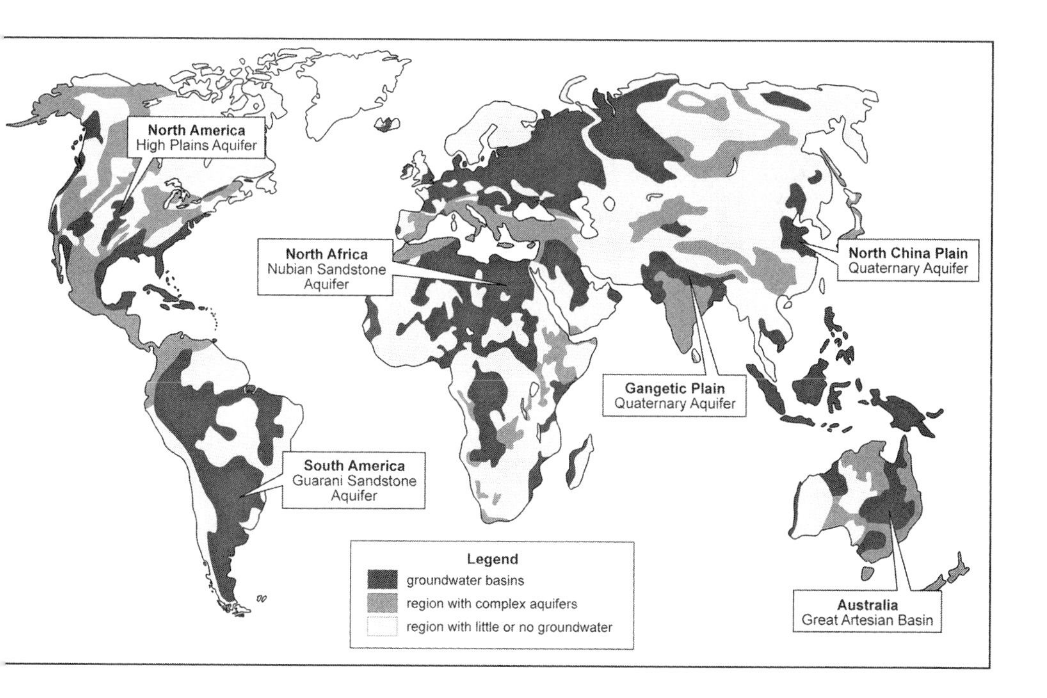

Simplified hydrogeological map of the world identifying some large aquifers discussed in this book. Local aquifers may provide important rural supplies in regions shown with little groundwater. Modified after S. S. D. Foster and P. J. Chilton, *Phil Trans Royal Society B* 358, no. 1440 (2003). Used with permission.

maybe for the first time in their lives, to start thinking about groundwater.[29]

This book takes it from there, giving the reader a breadth-and-depth appreciation of the main issues surrounding global groundwater use and abuse. Our goal is to provide an engaging and informative discussion about one of the world's most critical, but neglected, natural resources. We begin with three very different perspectives as the problem is unfolding in India, Arizona, and sub-Saharan Africa. We then examine the worldwide effects of groundwater use on aquifers, streams, land subsidence, and the environment, as well as how managed aquifer recharge and water recycling help counter these effects. Our discussion then turns to groundwater contamination, and the importance of protecting aquifers and wells from human-induced as well as naturally occurring contaminants. The final chapter summarizes key points essential for protecting and managing groundwater resources for current and future generations.

2

India's Silent Revolution

My grandfather once told me that there were two kinds of people:
those who do the work and those who take the credit.
He told me to try to be in the first group; there was much less competition.
—Indira Gandhi

July 31, 2012, began like any other day in India. The streets were choked with the usual assortment of cars, scooters, bicycles, rickshaws, and trucks and buses spewing clouds of diesel fumes. Pedestrians worked their way through the maze of street vendors and piles of reeking garbage. Radios blared. Horns honked madly. In the drenching July heat, the air was so foul you could cut it. It was just another day until five minutes after one, when the power went out.

Blackout Tuesday would make world history. But at first, people barely noticed. There had been a major power failure just the day before when India's northern grid had crashed. Rolling blackouts and localized outages, known as load shedding, are far from unusual—just another part of life as power grid controllers routinely cut power to keep the system running. Now, for the second day in a row it was time to crank up the diesel generator, if you were lucky enough to have one. Not even all hospitals in India have back-up generators. Everyone else just hunkered down and waited.

Minutes became hours. All across northern India thousands of stranded trains baked on the tracks. As the heat intensified throughout the afternoon, the sweating, dehydrated, exhausted people packed in those stranded trains wondered when the nightmare would end. Dead traffic lights turned the already choked streets into total gridlock. The daily business of modern life completely shut down. There was, however, one segment of the population that was only marginally affected, if at all. For the more than 300 million urban slum-dwellers and rural poor who have never had electricity, it was just another day.

The magnitude of the disaster slowly began to unfold. The blackout extended nearly two thousand miles (3,200 kilometers) from India's border with Pakistan to its opposite border with Myanmar, encompassing twenty of India's twenty-eight states—more than 75 percent of the country. Blackout Tuesday was the largest electrical blackout in history, affecting around 10 percent of the world's population.[1]

By evening, as power began to be restored, the government was squarely on the defensive. Top ministers scrambled for explanations. Why today? Why any day, for that matter? The grid, with all of its problems, had been limping along for decades.

Officials started pointing fingers. A former top electricity regulator explained that the national grid has a sophisticated system of circuit breakers that should have prevented such a massive blackout. The problem, he declared, is that elected state leaders, to keep themselves in office, demand that more power be diverted to their regions even if it threatens the national grid.[2]

He was just stating the obvious. As elsewhere in the world, corruption is rampant among Indian politicians. In 2008, nearly a quarter of the country's national parliament was facing criminal charges.[3] Yet there was still something upsetting about the power minister being promoted to one of the country's most important positions on Blackout Tuesday.

Others blamed electricity theft, referring to the rat's nests of wires on electrical poles that residents and businesses bribe someone to install, or just climb up and attach themselves. In some Indian states, power losses on high-voltage transmission lines are as high as 50 percent because of theft.[4]

Not overlooked in the blame game were the more than one hundred million farmers who, for decades, had been subsidized with virtually free electricity for pumping groundwater to irrigate their crops. In a drought year like this one, they were using even more power than usual.

In the eyes of many, the farmers were a huge part of the problem. They might even be the reason why the scales had finally tipped into a Blackout Tuesday.

India will soon become the world's most populated country, on a landmass about a third the size of China or the United States. All those people competing for breathing room, food, and resources translates into world-class problems. Out of the seemingly endless list of challenges, no other country in the world comes even close to pumping its groundwater as fast as India. The amount is a staggering 66 trillion gallons (250 trillion liters) a year, enough water to fill an eighteen-inch-diameter pipeline to the moon and back two thousand times.[5]

India is pumping more than a quarter of all countries in the world *combined*, and more than twice the amount pumped in the United States. The United States, which ranks close to China for the number two pumping slot, has a lot more groundwater to pump—if there weren't compelling reasons not to. As Henry Vaux, an expert on the economics of water use, explains, "persistent groundwater overdraft is always self-terminating." That is, there comes a point when it's no longer economically feasible to keep pumping water, which weighs about eight pounds a gallon, from greater and greater depths. But there's no such self-limiting effect in India, where many farmers don't have to worry about the cost of electricity.[6]

India's reliance on groundwater may seem particularly strange, given the country's abundant surface water. Some of the world's greatest rivers flow out of the Himalayas—the Yamuna, the Ganges, and the Brahmaputra. The explanation begins with climate. Much of the country is semi-arid, with almost all rain coming during the summer monsoon. In some places, the monsoon never comes. Much of the large state of Rajasthan, in the northwest, is bone-dry desert.

Then there's geology. Over two-thirds of India's landmass—all of peninsular India except for narrow coastal plains—is underlain by hard rock. Much is crystalline rock: granite and other igneous rocks that formed deep underground in a pressure cooker of intense heat. Hard rock, as the name suggests, is hard and very dense. Except for cracks and fissures, hopefully connected, water just doesn't get in. Consequently, two-thirds of India has very little storage capacity for groundwater.

The Deccan Traps, which span about a third of peninsular India, are another version of the same problem. The traps (Dutch for "steps") are a geologic wonder surpassed by few places on Earth. The action began around 70 million years ago, when the Indian plate passed over a hot spot deep in the mantle that began pumping magma up and out, resulting in one of the largest basalt flows in Earth's history. The plate stayed put for more than half a million years, with lava pouring out about every three hundred years. In this context, however, *poured* is an understatement. On the Deccan Traps, at times lava flowed at such an unfathomable rate that it would have filled about forty Olympic-sized swimming pools *every second.*[7] The result is a huge basalt plateau that is also hard rock.

Being doubly disadvantaged by climate and geology, it's reasonable to assume that peninsular India would be sparsely inhabited. The reality is just the opposite, with many millions of people living here. Cyberabad, the country's huge telecommunications network in Hyderabad and Bangladore, is in peninsular India. Severely water stressed and strained in all directions, over-pumping is definitely not a good idea. Yet peninsular India has a very large agricultural footprint supported by groundwater.

The events leading up to India's groundwater crisis began in the mid-1960s. The country was fifteen years into independence, with dreams of becoming an economic power (perhaps even a superpower), when drought hit. In India, droughts are fairly common and always a serious event. But this one was far from common, a perfect storm of an exploding population paired with crops dying in the fields for three consecutive years. The country faced the very real threat of mass starvation. India's leaders had no choice but to accept huge gifts of food grain from the United States and others. This state of affairs resulted in a national policy of achieving food security at any cost. This period also marked the beginning of the Green Revolution with its hybrid seeds, chemical fertilizers, and emphasis on irrigation.[8]

There had been efforts to promote groundwater irrigation since the 1930s, but the high initial investment of drilling a well, buying a pump, and installing power lines excluded most Indian farmers. During the drought crisis, state governments intensified their efforts by heavily subsidizing the initial capital outlay. This solved part of the problem, but drilling and hooking up a well is a one-time cost. Pumping water to the surface, day after day, requires a substantial ongoing investment that, again, excluded most farmers. It soon became apparent that the only way to get large numbers of farmers on board was to subsidize their electricity costs, as well as guarantee generous prices for rice and wheat.[9]

Within a few years, India had a booming agricultural economy based on groundwater. In 1950, the country had around 150,000 irrigation wells, with many of them using diesel fuel. By 1980, there were around one million electric-powered wells. In the mid-1990s, the number had climbed to somewhere around eight million. Five years later, there were twelve million. By 2010, sixteen million electric-powered wells (along with millions of diesel-powered wells) were pumping from falling water tables.[10]

The groundwater movement spread so quickly across India, as well as other parts of the developing world, that it became known as the Silent Revolution.[11] For many millions of farmers, pumping groundwater assured them of abundant crops even during times of drought. Once an impossible dream, many of these farmers were climbing out of extreme poverty. Some were moving into the middle class and able to educate their children.

The decision to subsidize electricity was born in a desperate time. It wasn't long before unintended consequences began to develop, particularly in northwestern and peninsular India where the electricity subsidies really took hold. (With little rural electrification and fewer government incentives, less groundwater is used in eastern India, where it is more plentiful.)

As the number of electric-powered wells grew, the State Electricity Boards (SEBs) were increasingly engaged in their new role in agricultural development. The boards have a monopoly over power generation, transmission, and distribution, so the burden fell on them to absorb the rising transaction costs of serving this new clientele. These costs were considerable, and included the usual things a power company does if it wants to get paid—installing meters, repairing and maintaining broken meters, reading meters, billing farmers, and collecting the money. And all of this over a vast countryside with poor roads and often inadequate bridges to reach isolated farming communities.

Soon there were other unanticipated problems, such as meter tampering and under billing by staff in collusion with farmers. Farmers were still a fairly small segment of the national power load, but these escalating transaction costs (not to mention outright fraud) were beginning to worry utility executives. A 1985 study by the Reserve Bank of India revealed that the transaction costs amounted to a third of each farmer's bill.[12]

Clearly, the situation had become untenable. After the usual meetings and poring over spreadsheets, the utilities decided to change the way that farmers were charged. Flat tariffs based on the pump's horsepower now replaced subsidized metered tariffs. The upside was no more legions of utility workers tromping over muddy roads to read meters. The downside was that it no longer mattered how much electricity farmers used—they just paid a set amount based on pump size.

Charging a flat fee for electricity is not exactly the way to encourage responsible power and water use. The flat tariff so drastically reduced the already marginal cost for pumping groundwater that it created an illusion that the power was free. This naturally increased demand for more wells and pumps. A growing feeling of entitlement to free power also led to viewing groundwater as a free and unlimited resource. India's long-held tradition of water conservation was replaced by wasteful use of its precious groundwater. By 2002, the agricultural sector was consuming around 30 percent of India's electricity. In peninsular India, the ratio approached 45 percent.[13]

In addition to this alarming drain on the power grid, groundwater was being pumped from increasing depths, often with electric pumps operating at around 40 percent efficiency. In many states, the true electricity cost for groundwater irrigation was anywhere from 50 to 100 percent of the value of the crop. In the southwestern state of Andhra Pradesh, the average farm income from groundwater irrigation was less than the cost to serve them power.[14] Many SEBs drifted into consistent annual losses that severely limited their ability to invest in new power plants and maintain transmission and distribution lines. In an attempt to stay fiscally afloat, the SEBs began to pass off their losses to nonfarm customers.

At this point, the nonfarming public was getting upset. On top of having to deal with voltage fluctuations and power interruptions, *and* getting stuck with much of the farmers' electricity tab, they were becoming increasingly concerned about their water supply. At least 85 percent of India's rural population depends on groundwater for their daily needs.[15] As farmers continued to pump down the water table, villagers who depended on their well for domestic use began to find themselves without water.

Groundwater is viewed as a democratic resource, but as water tables decline, the poorest people lose out first as their shallow hand-dug wells go dry. Continuing declines increasingly affect small and marginal farmers. Eventually only the wealthier farmers have the capital to drill deeper wells and buy ever more powerful pumps. Those no longer able to make a living from farming are forced to rent or sell their land.[16]

For decades, Tushaar Shah, a senior fellow at the International Water Management Institute based in Sri Lanka, has been trying to reason with politicians, farmers, and government officials about this "massive water-scavenging free-for-all," as he puts it. As a highly respected expert on India's water problems, Shah minces no words about the seriousness of the problem. "When the balloon bursts," he warns, "untold anarchy will be the lot of rural India."[17] It's not just villagers who are at risk. Nearly half the urban water supply comes from groundwater. All told, drinking water for nearly a billion Indians is potentially at risk.

With increasing numbers of people worried about their water supply, a push was on for the government to *do something* about the way electricity is billed to farmers (who comprised about 65 percent of

India's population). This initiative soon got the farmers' attention, and they did something. With such a potential and formidable lobby, they began to organize into powerful "vote banks." By the 1980s, farm power pricing and supply policies were virtually in the hands of the farmers. Anyone trying to get into (or stay in) political office had to bow to the will of the farmers or commit political suicide.

At this point, there was still one thing the State Electricity Boards could do—and they did it. The SEBs began reducing the daily hours that the farmers' three-phase power was available, while maintaining single- or two-phase power for domestic users. Once again, the plan backfired. The frequent and unpredictable power interruptions led to farmers leaving their pumps switched on to pump whenever power was available, whether the water was needed or not. The practice of stealing electricity also began to take hold.[18]

Perhaps nowhere in the world do farmers have the same power to topple their national government as in India. In the early 2000s, the national government was getting top marks on the macroeconomic front—high GDP growth and high foreign institutional investment—but its members were voted out of office because they were considering electricity reforms. State ministers didn't make the same mistake. The new chief minister of Andhra Pradesh promised free electricity to farmers on the day he took office. The chief minister of neighboring Tamil Nadu followed suit.[19]

In 2005, the World Bank warned that without significant changes in its water resource management, India's demand for water will exceed all sources of supply by 2020. Meanwhile, India's unsustainable policies for pumping groundwater continue unchecked.[20]

In the late 1980s, a groundwater recharging movement took hold in Mahatma Gandhi's native state of Gujarat.[21] About a third of Gujarat is taken up by the region of Saurashtra. Like much of India, Saurashtra has a semi-arid climate and is underlain by hard basaltic rock with very limited groundwater storage capacity. In addition, Saurashtra is highly vulnerable to droughts.

During a severe three-year drought in the mid-1980s, almost all of Saurashtra's wells dried up. Due to the region's clay soils, when the monsoon rains finally returned, most rainwater just ran into seasonal streams and the wells remained dry. In a desperate attempt to recharge their wells, a few farmers began channeling monsoonal runoff into them. This was the exact opposite of the usual practice of diverting the runoff to a neighbor's field or common land. Other farmers took note and began recharging their own wells. Gradually, this individualized well-recharging movement evolved into a widespread groundwater recharging movement. The people of Saurashtra began building water-harvesting structures, such as check dams or farm ponds. Between 1992 and 1996, more than ninety thousand wells were recharged in Saurashtra, and hundreds of farm ponds were constructed for groundwater recharge. In just four years, the movement spread to hundreds of villages and benefited thousands of families.

Harvesting rainwater for recharge was not a new idea, but what makes Saurashtra unique is the way the movement spread. A few farmers recharging their wells with rainwater would have had little impact, but when entire communities joined together, the effects on groundwater levels and crop yields were significant. With increasing numbers of farmers practicing water harvesting, it made sense for neighboring farmers to follow suit.

Saurashtra's groundwater recharging movement was inspired by the Hindu sect Swadhyaya Pariwar. *Swadhyaya* is a Sanskrit term meaning "the study of self or introspection"; *pariwar* means "family." The Swadhyayees practice devotion to God by giving to the community in a selfless way. This includes creating "common properties" from devotional offerings. Many Swadhyayee coastal villages in Saurashtra set up fishing boats that are owned by the entire village and managed by voluntary labor. These community boats donate their catch to widows, the poor, the disabled, and the elderly as a village offering to God.[22]

Pandurang Shastri Athavale, a charismatic teacher who espoused a philosophy of life based on the *Bhagavad Gita*, led the initial groundwater recharging movement. Athavale began asking his followers why farmers in Saurashtra could not adopt the rain harvesting and conservation techniques used in Israel and elsewhere. Addressing huge crowds, often between 100,000 and 200,000 people, he exhorted them to develop innovative ways of conserving water. "If you quench the thirst of Mother Earth, she will quench yours," he told his followers.[23]

Efforts to organize the recharge movement came through the Saurashtra People's Platform, which compiled information about groundwater recharge technologies and their impact. Group members communicated this information to the generally illiterate farmers by mass-distributing leaflets with easy-to-follow illustrations. The groundwater recharge movement spread beyond the Swadhyayees to farmers of all religious and political persuasions.[24]

Gujarat's groundwater recharging initiatives became a large-scale and self-propagating movement that was sustained by its own spirit and energy. All this effort, however, addressed only part of the problem—increasing water supply. Much more challenging was reducing the demand created by the excessive groundwater pumping tied to farmers' subsidized electricity. In 2003, Gujarat began a landmark program to provide dependable power to villages, while at the same time rationing electricity to the farmers. The success of this program rests, in part, on the leadership of Gujarat's new chief minister, Narendra Modi.

As a Hindu Nationalist, Modi is a controversial figure, but he has proved to be an effective leader in many ways. After being in office for just over a year, Modi took on the electricity and groundwater problem. Through the newly conceived proposal called Jyotigram Yojana (Village of Light), farmers would be given dependable (but rationed) electricity, while continuous power would be supplied to all eighteen thousand villages in Gujarat. People scoffed. It was simply unimaginable to have complete rural electrification in the state, not to mention quality power. And where was the money going to come from? Like every other SEB in India, the Gujarat Electrical Board was just managing to stay above financial ruin.

Modi and his energy minister faced the problem head on. Heavily subsidized electricity to farmers had led to critical levels of overpumping, which in turn had led to a nearly bankrupt electrical industry. Sporadic power to rural communities was exacerbating poverty and curtailing self-initiative and economic opportunity. Viewed from a larger context, food production would increasingly be hampered by the scarcity of water. Something had to be done before the crisis spiraled beyond any solution.

Under the Jyotigram Yojana program, farmers would receive a full-voltage, uninterrupted power supply. In return, they would pay a fair, yet still subsidized, price for electricity. Power also would be rationed to eight hours a day. Making this plan a reality would require completely rewiring Gujarat.

The proposal set off the usual farmer opposition, but the completely unexpected happened when the government didn't blink or back down. Eventually the farmers recognized the advantages of paying for a reliable, quality power supply. Jyotigram Yojana was launched in September 2003. The state erected 1.2 million power poles, installed 12,600 transformers, and strung 30,000 miles (49,000 kilometers) of electric lines.[25] In just two and a half years, Gujarat accomplished the seemingly impossible—separate power lines for farming and nonfarming uses. For the first time, farmers were forced to use electricity and groundwater more efficiently, while Gujarat's villages had uninterrupted power.

By 2012, Gujarat had a profitable state electricity board and a rising water table. Jyotigram Yojana had radically improved the quality of village life, spurred nonfarm economic enterprises, halved the power subsidy for farmers, and led to agricultural growth. Gujarat became the only state in India with an energy surplus. The vision, strategic planning, and commitment demonstrated in Gujarat has become a model for India's national government.[26]

In 2014, Narendra Modi was elected prime minister of India in a landslide victory. Modi has resurrected concepts for a massive interbasin transfer project (similar to China's South-North Water Transfer Project) to move water from the Himalayas and other areas of surplus to water-short areas of India. The Modi government also plans to replace (for free) twenty million electric-powered pumps used by farmers with more efficient models. Even if feasible, these are only partial solutions to India's groundwater-energy dilemma, particularly in the face of climate change and a growing population. Although India is the most extreme case, it is not alone. Electricity subsidies continue to be a major factor leading to severe groundwater declines in Pakistan, Mexico, and elsewhere.[27]

3

Arizona

And it never failed that during the dry years the people forgot about the rich years, and during the wet years they lost all memory of the dry years. It was always that way.

—John Steinbeck

The 185 miles (three hundred kilometers) of interstate highway from Gallup to Flagstaff is known as the Speedway for good reason. In the summer months, with the sun climbing toward the sky's zenith, it's best not to find yourself on this wind-blasted desert after midmorning. Even in an air-conditioned car you will feel the heat as a monstrous presence. Foot down, pedal to the metal, the goal is to get across the Speedway as quickly as possible. It's a good idea to stay hydrated, just in case. Speaking of water, there is none, until you pass the mostly dry channel of the Little Colorado River meandering north to connect up with its big brother in the Grand Canyon.

Gradually the red-rock country transitions through grassy plains to the tree-covered plateau of Flagstaff. At seven thousand feet (2,100 meters), Flagstaff is one of the highest cities in the United States. Northwest of Flagstaff loom the San Francisco Peaks where trails, streams, and deer wander through lush meadows and forests of ponderosa pine, oak, and aspen. Summers here are reportedly perfect. Most cars turn south and head over the edge of the Mogollon Rim.

The Rim is a major flora and fauna boundary between the mile-high Colorado Plateau and the Central Highlands. The road is steep, spectacular, and definitely not a speedway. Suddenly, there's too much to see: canyons of pink and blonde rock, sharply pinnacled ridges, and blue sky shimmering iridescent in the bone-dry air. Eagles, hawks, and ravens soar the thermals, on the hunt for anything that dares to even think about moving down below. The Central Highlands is world-class rock country. Sedona is off to the west.

Suddenly, the first saguaros appear. A mile or two further these majestic giants are everywhere, spaced at regular intervals as if someone had taken a yardstick and carefully laid them out. We're not just talking about a high water I.Q. These behemoths are water geniuses. Spread out! Crowding strictly prohibited. When it rains, these living reservoirs suck up their every allotted drop, storing it in their massive trunk and limbs, for years (if necessary) of drought and blast furnace heat. The saguaros inform you that you're down, that you're back on brutal desert, and that it's possible to survive here if you're very, *very* smart about water.

It's now around three hundred miles (five hundred kilometers) since entering Arizona way back up there to the northeast. Over all that distance there have been turnoffs for an occasional small town and one modest-sized city. The norm, in this desert state, is desert. Until now. Like the saguaros, houses are suddenly everywhere. But unlike the saguaros, they're jam-packed together. And that scenic two-lane highway has suddenly morphed into a major freeway with the standard hustle and tangle of traffic.

Welcome to Phoenix.

There's still a desert out there somewhere, but here it's been erased and replaced by metropolitan sprawl. This could almost be L.A., except for the highway beautification. Mile after mile of exquisite native plants in beds of red gravel are set off against walls of stunning mosaic. Overpasses and their sweeping entrance and exit ramps have been transformed into soaring compositions of geometric design, colored in soft desert pink and the deep maroon of the volcanic mountains to the east. Could all this xeriscaping, elevated to high art, be some kind of message to the multitudes? Perhaps an attempt to educate the 4.3 million people living here that look what we did (and you can *too*) without water.

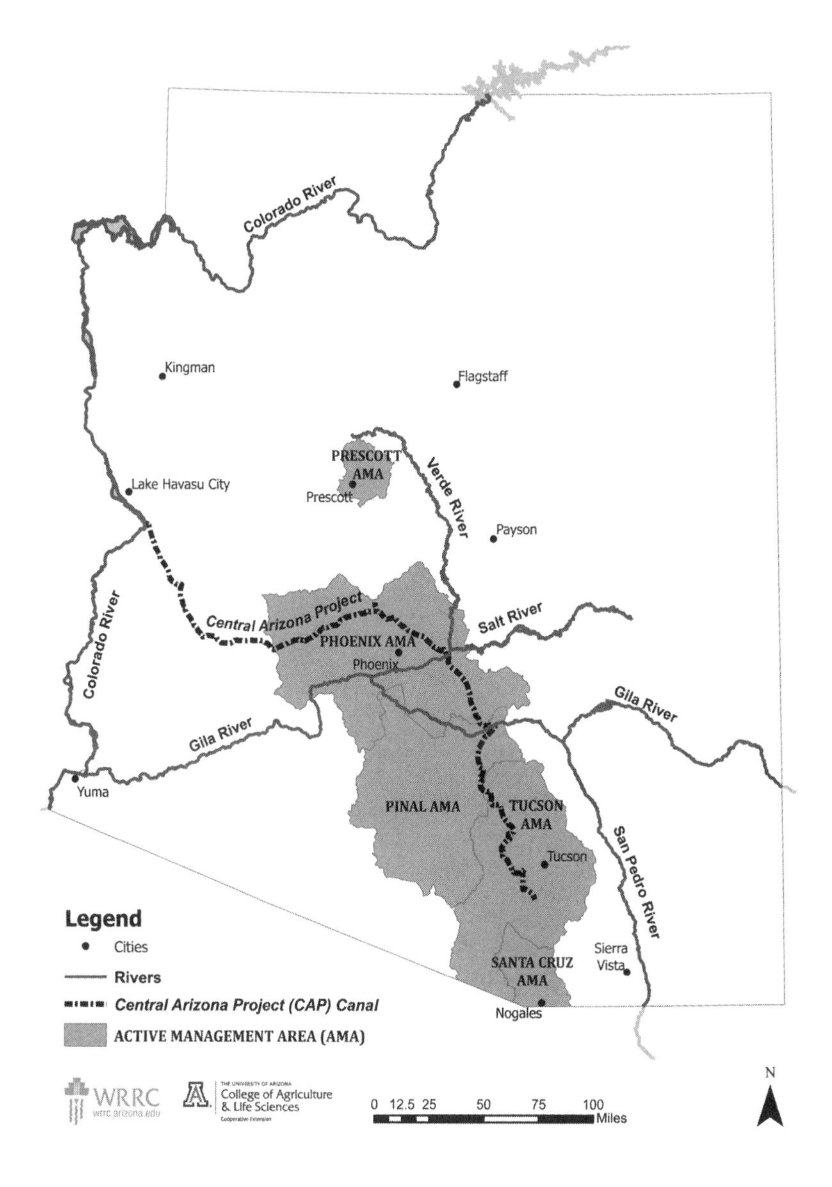

Active Management Areas (AMAs), Central Arizona Project canal and selected rivers in Arizona. *Source:* Water Resources Research Center, College of Agriculture and Life Sciences, University of Arizona. Used with permission.

The message doesn't appear to be sinking in. Phoenicians, it turns out, love grass. Many residential neighborhoods are full of the stuff. Among the world's major cities, only a few—such as Baghdad, Cairo, and Riyadh—rival Phoenix for less rain. In the United States, Phoenix is the hottest and driest major city with a meager seven inches (180 millimeters) of rain a year—and the people living here are watering grass. But that's just the beginning. Phoenix has around 250 golf courses. And then there are the water parks. Leading the pack is Wet'n'Wild with over thirty water attractions such as water slides, water rafting, and a giant wave pool. Golfland Sunsplash, another water extravaganza, has the usual assortment of water slides and wave pool, plus an "endless river" for an afternoon of relaxed rafting. Phoenix also has six hotel resorts with their own water playgrounds. The Westin Kierland Resort and Spa boasts a nine-hundred-foot lazy river and a FlowRider that allows visitors to surf on the desert. As they told us at the Visit Phoenix Information Center, "You'd be surprised how much water we splash around here on the Sonoran Desert!"

Nonetheless, Phoenix has come a long way in managing its water. Virtually all of its wastewater is recycled to supply cooling water for the country's largest power plant and for watering golf courses, school lawns, and other public places.

South of Phoenix the desert continues. Once again, sparse desert plants are armed with thorns or dagger-tipped leaves (or both) to keep animals from breaking in and stealing their carefully hoarded water. Against this brutal desert backdrop, suddenly lush green fields begin to appear as far as the eye can see—farming on the grand scale. Irrespective of how out-of-place it looks, this desert state has a large agricultural economy. The rich soils and arid, bug-free climate provide two of the three requirements for bountiful crops. The serious lack of rain kept this region from realizing its agrarian potential until the centrifugal pump arrived and Arizona's farmers began large-scale pumping of groundwater. It wasn't long before much of the desert around and between Phoenix and Tucson was under cultivation.

Geologically, this is the basin and range province of the United States and Mexico, where north-south trending mountain ranges are separated by nearly flat desert basins. These basins hold vast amounts of groundwater—much of it a legacy from the colder and wetter climates of the last ice age. With minimal opportunities for recharge, this fossil water is basically a nonrenewable resource. Nonetheless, once armed with the centrifugal pump, Arizona's farmers began extracting it like there was no tomorrow. The mining industry also began pumping large amounts of groundwater. In the 1960s, Arizona became the second-fastest-growing state, with around 80 percent of the growth occurring in Phoenix and Tucson. All that fossil groundwater helped make it possible.

Phoenix and Tucson have very different water situations. Tucson, the Old Pueblo, dates back to Spanish colonial times. Early inhabitants were amply supplied with water from the Santa Cruz River and the springs near Mission San Xavier. As farms and communities increased in number and size, the springs dried up and the river could no longer supply the necessary water. When the centrifugal pump came along, groundwater levels started dropping fast. There were other consequences. Groundwater and surface water are dynamically interconnected. After groundwater levels began to fall, it wasn't long before the Santa Cruz River completely dried up. By 1980, Tucson had become one of the largest metropolitan areas in the world that was completely dependent on groundwater. The city had no choice but to continue rapidly "mining" its precious fossil water endowment.

That wasn't the end of it. In some geologic settings, overpumping groundwater causes the land surface to drop. Land subsidence first appeared in the agricultural areas near Tucson and Phoenix. By 1980, with groundwater levels continuing to plummet, more than three thousand square miles (7,800 square kilometers) had been affected by subsidence—including parts of metropolitan Tucson and Phoenix.[1] A dropping land surface in a city can wreak havoc with foundations, water and sewer lines, roads, and sidewalks. Subsidence can be slowed or halted by limiting or stopping pumping, but prior land subsidence is mostly irreversible.

There was yet another problem, as earth fissures began forming along the edges of some of these subsiding basins. Fissures first appear as barely noticeable cracks on the land surface. If pumping continues, the fissure keeps expanding. Some grew to thousands of feet long and became gaping ravines more than thirty feet (nine meters) deep. Earth fissures can (and do) seriously damage highways, railroads, sewer and water lines, buildings, and anything else in their way. Needless to say, Tucsonans were becoming desperate for another source of water.[2]

Phoenix also was pumping large amounts of groundwater and having its own problems with land subsidence, but unlike Tucson, it has a generous supply of surface water from the Salt River and its tributaries flowing out of the mountains. Phoenix also has the Salt River Project, consisting of seven dams and over 130 miles of canals that deliver water to the city. The Salt River Project moved into high gear with the completion of Roosevelt Dam in 1911—a year before Arizona became the forty-eighth state. The tallest masonry dam ever built, Roosevelt Dam was considered a technological marvel at the time. President Theodore Roosevelt (the dam's namesake) journeyed to this outpost of civilization in order to speak at the dedication. Roosevelt told the audience that, with this new water supply, he could foresee a day when the Phoenix Valley might hold 150,000 people.[3] By the early 1970s, the Phoenix metropolitan area had grown to more than a million people. Like Tucson, they needed another source of water.

All eyes turned to the Colorado River.

This wasn't the first time that Arizonans had looked to the Colorado River to solve their water shortage, but there were longstanding obstacles to getting that water. First and foremost, the river is located more than two hundred miles (320 kilometers) from the population and agricultural heartland in the center of the state. Nonetheless, since the days of Harry Truman, Arizona had lobbied Congress for what they called the Central Arizona Project—an extensive canal system that would bring river water to where it was needed. Congress repeatedly said no.

When it comes to an engineering challenge, when there's a will, there's usually a way. So that wasn't the problem. Beginning with Hoover Dam in the 1930s, Congress had been picking up the tab on

The Picacho earth fissure located between Phoenix and Tucson is ten miles long.
Source: U.S. Geological Survey Circular 1186.

any number of big water projects with staggering price tags, so money wasn't a game-stopper either. The only real obstacle keeping Arizona from getting its fair share of Colorado River water was one of geography. The river forms the boundary between Arizona and California, and Californians saw no good reason to give up so much as a drop of river water—even if some of it didn't legally belong to them. This was a full-fledged water war, and it had been going on for decades.

California contributes virtually nothing to the flow of the river; by contrast, every river system in Arizona drains into the Colorado. Yet the agricultural potential of California's southeastern region resulted in it becoming the first major user of Colorado River water. This part of the state was known as the Colorado Desert until 1900, when land developer George Chaffey sought to attract settlers by renaming this inhospitable, but immensely fertile, region the Imperial Valley. The ad campaign worked. This desert, with an annual rainfall of just three inches (seventy-six millimeters), eventually became one of the most productive farming regions in the country.[4]

Initial diversions from the river to the Imperial Valley began in 1901. Soon thereafter, a second canal was cut south of the Mexico border by an unscrupulous group trying to avoid any control by the U.S. government. Disaster struck when a surge of floodwater took out the shoddily constructed new diversion. Within days, the entire Colorado River was pouring through a mile-wide break into the Imperial Valley—destroying buildings, drowning crops, and creating today's Salton Sea. It took two years of herculean efforts and millions of dollars to get the river back in its channel. In the aftermath, California began to petition Congress for flood control on the Colorado River and an All-American Canal to bring water to the Imperial Valley without passing into Mexico. For Arizonans, this was an early shot across the bow from their water thirsty neighbor to the west.

Despite its large tributary system, the Colorado River ranks only sixth in volume among major rivers in the United States. The land it drains, however, is some of the most arid and hostile terrain in North America, making the Colorado River the main lifeline for survival. It was only a matter of time before the seven basin states that had claim on the river realized that they needed to come up with an agreement for who gets how much.

Though far from easy, the first step was the easiest. In 1922, representatives from all seven states signed the Colorado River Compact, which divided the river into two parts. Colorado, Wyoming, Utah, and New Mexico became the Upper Basin. Nevada, Arizona, and California became the Lower Basin. Each basin would get an annual allotment of 7.5 million acre-feet, with one acre-foot being equal to the volume of water needed to cover an acre of land to the depth of one foot. The compact also stipulated that the Lower Basin could increase its share by a million acre-feet (1.2 billion cubic meters) from tributaries to the lower Colorado. All the representatives signed and went home to their respective states to work for ratification.

The Arizona legislature had two big problems with ratifying the compact. First, it didn't specify how much water each state would get; that detail needed to be worked out among the states in each basin. Second, the compact didn't state it directly, but the Lower Basin's "additional" million acre-feet would largely come from Arizona's Gila River. The bottom line was that Arizona not only had to work out its allotment with California, but that water behemoth to the west was claiming that the Gila River counted as part of Arizona's allotment. There was little Arizonans could do about it. California had the population, wealth, and political muscle to run right over the much smaller and sparsely populated state of Arizona. It wasn't long before it did just that.

In 1924, William Mulholland—Los Angeles's infamous water czar and one of the most powerful men in the state—appeared before Congress. Ten years earlier, Mulholland had masterminded diverting the main water supply from the farmers and ranchers in the Owens Valley. Now Mulholland was telling Congress that Los Angeles needed more water. Congress listened.[5]

None of the states were completely satisfied with the compact's terms, but no one wanted to rock the boat. The Upper Basin states feared that without a compact, California would take as much water as it wanted. Meanwhile, California needed to get the compact ratified to pave the way for its dream of a super dam to supply the state with enough water for future growth. When Congress announced that only six of the seven states would need to ratify the Colorado River compact, Arizona was out-maneuvered and isolated.[6]

Arizona's leaders fought bitterly, but in 1928 Congress passed the Boulder Canyon Project Act, which codified the Colorado River Compact and authorized construction of Boulder (later renamed Hoover) Dam. The act also authorized the All-American Canal and Imperial Dam for the water-thirsty Imperial Valley farmers, and established the annual allotments for the Lower Basin states—California garnered 4.4 million acre-feet, Arizona got 2.8 million acre-feet, and Nevada received a paltry 300,000 acre-feet. Finally, California and Arizona would split any "excess" water in the Lower Basin. With Arizona having no way of using anywhere near its full share, the door was open for California to take much more than its entitlement.

In 1931, in the midst of the Great Depression and while the great dam was under construction, southern California voters approved bonds equivalent to more than 10 percent of the region's entire property value to finance pumps, pipelines, tunnels, and aqueducts to lift water from the Colorado River and transport it more than two hundred miles (320 kilometers) to the California coast.[7] Soon thereafter, construction began on Parker Dam, which would divert much of the river into the aqueduct.

For Arizonans, this was the last straw. The governor dispatched National Guard and militia units to prevent any attempt to anchor the dam to the Arizona side of the river. This was the closest that any two states had come to armed conflict since the Civil War. Only half in jest, the *Los Angeles Times* sent a "war correspondent" to the site. Construction was halted until Congress officially sanctioned the project.[8]

Arizona's only hope was to fight through the courts. Throughout the 1930s, the state was a party to four U.S. Supreme Court cases contesting the Colorado River allotments. It lost every case.[9]

In 1940, Arizonans elected Governor Sidney P. Osborn, who steered the state in a new direction. Osborn had long been one of the fiercest champions in the fight with California for more water, but as governor, he decided it was time to change tactics. "Whatever our previous opinions," he told his fellow Arizonans, "we now can only recognize that the decisions have been made."[10] He believed that it was time to accept that California had a much larger allotment than Arizona, and there wasn't anything they could do about it. Osborn argued that the state's new focus should be on getting the water to which Arizonans were entitled. At long last, in 1944 the Arizona legislature finally ratified the Colorado River Compact. For the first time, California faced a real possibility of losing Arizona's excess water.

Arizona still had to clear three huge hurdles with Congress before there would be any action on the Central Arizona Project (CAP). First, California and Arizona needed to settle the question of the Gila River. Second, Arizona needed to pass a groundwater law to limit pumping in areas that would be served by the CAP. After all, why should the federal government come to the state's rescue with a huge federal project if the state couldn't manage what they had? Third, and the only hurdle that seemed surmountable, Arizona needed to figure out how it would repay construction costs and oversee management of the project.[11]

In 1952, Los Angeles installed a second battery of pumps at the head of its aqueduct at Parker Dam. The state's diversion now climbed toward 5.3 million acre-feet—900,000 more than its entitlement and just shy of taking *all* of Arizona's "excess" water. In desperation, Arizona once again went to the Supreme Court.[12]

Arizona v. California was one of the longest, costliest, and most hotly contested cases in U.S. Supreme Court history. In addition, the trial's political and economic consequences make it the single most important water litigation case in U.S. history. The Court appointed a "special master"—Simon Rifkind, a New York lawyer—to run the case. With more than fifty lawyers and 340 witnesses testifying, and with thousands of exhibits received and a court transcript that eventually filled twenty-five thousand pages, this was a herculean task. Rifkind suffered a heart attack halfway through the eleven-year trial. California had a vested interest in dragging the case out, because every year the trial continued, the state got an additional three hundred billion gallons (1,100 billion liters) of water. California's chief attorney perfected the art of "dilatory obfuscation" by taking days to cross-examine a witness over something that could have been settled in five minutes or less. Rifkind recovered from his heart attack and finished the job, eventually turning in a 433-page report of his conclusions.[13]

In 1963, the Supreme Court finally ruled. To California's utter astonishment, the Court upheld Arizona's position on virtually all counts. Arizona was entitled to 2.8 million acre-feet of Colorado River water, and the Gila River didn't count one drop against its allotment.

There was one last hurdle before Congress would fund the Central Arizona Project. It may not have looked like such a big deal on paper, but passing a groundwater law in a state where farmers ruled the roost, and had always pumped as much as they wanted, was going to be an uphill battle.

For decades, Kathleen Ferris has been a key player in tackling Arizona's water problems. "I just got into water by accident," she told us after we settled into her office in downtown Phoenix. During Ferris's third year of law school, she completed an internship with Governor Calvin Rampton of Utah. She needed to pick an issue to focus on, discovered the governor was interested in water, and decided that water sounded interesting. After a brief period on Rampton's staff, Ferris moved to Arizona in 1977. For someone right out of law school with an interest in water, Ferris couldn't have moved to Arizona at a more tumultuous time. The Arizona Supreme Court had just ruled on a case with major repercussions for cities and mines throughout the state.[14]

For decades, Tucson had been reaching out farther and farther for water by purchasing land just for the water rights, then drilling wells and pumping the groundwater back to the city. Buying land solely for the water rights had become such standard practice that it had earned the folksy sobriquet "water ranching." Everything changed when the city and several copper mines began pumping near a large corporate pecan farm. The owner, FICO, went to court claiming that the farm was being damaged by the nearby pumping. The court ruled in FICO's favor, but only because water was being transported. Under Arizona's reasonable use doctrine, as long as a landowner withdraws groundwater in order to make reasonable use of his or her property, neighboring landowners have no claim for damages—even if the withdrawals have adversely affected groundwater levels under their property. Transporting water for use elsewhere was another story.[15]

The ruling was deeply unsettling to Arizona's growing cities and huge mining industry, but for Tucson its lifeline had been severed. Being completely dependent on groundwater, where was the city going to get water if it couldn't pipe it in from somewhere else? Arizona's cities and mines banded together against the farmers, descended on the state capital, and demanded some kind of legislative fix to this intolerable situation. Ferris arrived in Phoenix when the legislature was looking for someone who had some experience in water. "I didn't have much," she told us, "but far more than just about anybody else they interviewed."

Alfredo Gutierrez, the majority leader of the state senate, made Ferris his staff person for this complex and highly charged legal battle. For three months, a group of high-priced lobbyists tried to hammer out a compromise. All they accomplished was passing some temporary legislation that allowed Tucson and the mines to continue pumping until the whole mess could be sorted out and fixed. The group also established a Groundwater Management Study Commission, and gave it two and a half years to do the sorting and fixing by way of drafting a comprehensive groundwater law for Arizona. Ferris applied for the job of executive director of the commission. "Most people didn't want it," she laughed, "but I was young and naïve." She got the job.

This was the fourth groundwater management study since 1940. There weren't a lot of expectations. "It's very easy to kill water legislation," Ferris explained. "You'd have people reach an agreement, then someone at the legislative level would say, 'Well, I don't like that' and kill it." Yet Ferris's commission was different in a key way that gave it a fighting chance—a majority of the members were state senators or representatives. For the first time, the state had a groundwater commission with some political teeth.

The Groundwater Management Study Commission began in 1977, around the same time that newly elected president Jimmy Carter issued a hit list of nineteen water projects that he considered wasteful boondoggles. Carter promised to veto any legislation funding these projects. The Central Arizona Project was on the list. Although Carter soon backed down, this temporary crisis was a wake-up call for Arizona. The state needed to get a groundwater management law passed and get that canal built before the federal government changed its mind.

The turning point came when Bruce Babbitt became governor in 1978. "Bruce was very interested in water," Ferris explained, "and he's very, very smart." Early on in the commission's work, Ferris got to know him and kept him up-to-date on what was going on. She also developed good relationships with key staff members of Secretary of the Interior Cecil Andrus. As gatekeeper to Congress's purse strings, Andrus was critical to the cause. The commission needed him solidly on board.

After nearly a year of work, the commission's municipal and mining representatives drafted a plan for legislation that was staunchly opposed by agriculture. For the first time in Arizona's history, agriculture was outvoted and the cities and mines had the upper hand. Everyone recognized, however, that without the farmers' support, the chances of getting something through the legislature were slim to nothing.

Another year of work resulted in the possibility of a breakthrough, but the commission was also getting uncomfortably close to its two-and-a-half-year deadline. "By the time we'd gotten there," Ferris told us, "these guys, they had put in a lot of time and energy on this. They had become pretty personally invested in getting something done." As proof of this sincerity, the twenty-five-member commission came to a surprising decision. Realizing that there were too many cooks in the kitchen, they volunteered to cut their numbers. They then invited Governor Babbitt to come in and mediate the negotiations. After daily marathon sessions, Ferris met with Babbitt every evening to figure out what he wanted to tackle the next day. They also worked with the Interior Department, so there was a lot of back and forth between Ferris and the staff at Interior, as well as between Babbitt and Andrus.

Secretary Andrus and Governor Babbitt were good friends. They were both staunch conservationists who recognized that Arizona's groundwater overdraft problem was threatening to become a full-scale disaster. No one knew who had the rights to what. There were no limits on who could put in wells, and how much they could pump. The state's economic and population growth depended on getting this problem under control. Andrus and Babbitt decided to use the Central Arizona Project as the big stick to get the job done.

They cooked up a "scheme," as Ferris put it. It was the classic good guy versus bad guy foil. Secretary Andrus, as the apparent bad guy, began to deliver statements (written with the help of Babbitt's staff) that the CAP would not be funded if they didn't get a groundwater pumping law passed. Governor Babbitt, the good guy, would then protest that "if we don't do what Secretary Andrus wants, we won't get the Central Arizona Project."

"It was fun," Ferris laughed. "At one point, Andrus came to town and laid down the law and said, you guys have to do something in six months or we're going to cut off funding to the CAP. We had a closed door meeting with around 50 people, representing all the major water using interests. Andrus came in with this prepared speech and just laid it all down, and people were just like . . . WHOA . . . how can you threaten us and all that kind of stuff. There were articles in the newspapers about it. It was very public. And it kept everyone at the table. That's when we did six months of marathon negotiations. I was much younger then, thank God. I never would have been able to do it now."

With the June deadline looming, the job of pulling it all together into a solid legal framework fell on the shoulders of Kathleen Ferris—a twenty-nine-year-old just a few years out of law school. "At one point, it's like eight o'clock at night and everybody comes down to my grim little office in the basement of the Senate, and all the lobbyists and the commission members are saying, 'You've got to do it. You've got to do it,' and I remember bursting out, 'I can't do this!' I mean, we had been working night and day, and I thought there's no way we're going to get this done. But we did. We did it."

In 1980, after two-and-a-half years of intense negotiations, Governor Bruce Babbitt signed the Groundwater Management Act into law. This visionary approach to groundwater management continues to be unique in the United States, as well as the rest of the world. The act is a hard one to follow.

The act specified four Active Management Areas (AMAs) that cover much of the state's population and groundwater use, including large areas around Tucson and Phoenix, the agricultural region between them, and the Prescott area in the Central Highlands. (A fifth AMA was later carved out of the southern part of the Tucson AMA.) Within the AMAs, all large users of groundwater were now regulated. Meters were installed on the pumps. The act created the Arizona Department of Water Resources to monitor compliance, and gave it the power to enforce the rules. And all this in Arizona—a state that has fits over even the *thought* of regulation.

The act requires the development of progressive ten-year management plans, with mandatory conservation measures on major users in order to meet the long-term goal for each AMA. The goal for the Phoenix, Tucson, and Prescott AMAs is to achieve safe yield (defined as a balance between pumping and recharge) by 2025. "There's no penalty for failing to meet these goals, but they're working in good faith to get there," Ferris told us.

The act's most radical provision was that new residential subdivisions require a proven one-hundred-year "assured water supply": that is, a source of water must be physically, continuously, and legally available for the next hundred years. The water must come primarily from renewable water supplies, such as Colorado River water or treated wastewater.[16]

With most of the state's groundwater going to agriculture, and an eye toward future population growth, no new land in the AMAs is allowed to come under irrigation. (The same rule applies to designated INAs—Irrigation Nonexpansion Areas outside the AMAs—that are dealing with their own serious overpumping problems.) Owners of large wells are required to meter their groundwater pumping and report withdrawals. Given the state's continuing population growth, farmers are required to implement conservation measures so there will be groundwater left for eventual development.[17]

After a six-decade water war with California, the Groundwater Management Act completed the state's last hurdle and Congress finally authorized funding for the Central Arizona Project. But there was a catch. To obtain California's powerful congressional support for the project, Arizona was forced to accept junior priority status for CAP water from the Colorado River. In other words, if a shortage is declared on the lower Colorado River, Arizona could lose its entire CAP allocation before California loses any water.

The Central Arizona Project is the largest and most expensive water conveyance project ever constructed in the United States. The canal stretches 336 miles (540 kilometers), from Lake Havasu to southwest of Tucson. Every year, five hundred billion gallons of water flow through the canal. Fourteen pumping stations along the way lift the water almost three thousand feet (nine hundred meters), which translates into 2.8 billion kilowatt hours of energy required to lift the 4.5 trillion pounds of water moving through the canal each year. The price tag for constructing the Central Arizona Project was about $4 billion.[18]

The economic benefits of the CAP have been huge, approaching $100 billion a year in Arizona's gross state product.[19] There were also unforeseen problems. No one had done a study to see if the farmers would be willing to pay for the cost of CAP water. Arizona's farmers already had a water source, which enabled them to bide their time at the bargaining table until they got their price. Meanwhile, state officials couldn't afford to wait. They needed to sell the water so they could start paying back the federal government. To resolve the problem, farmers were given discounts on CAP water. They are also the first to be cut in the event of a Colorado River shortage.

Another unforeseen problem played out in Tucson. Completion of this leg of the canal was accompanied by media fanfare and a well-publicized taste test. Water officials were aware that CAP water contains more dissolved solids and has a different chemistry than Tucson's groundwater, so there were warnings for kidney dialysis patients, people on restricted salt diets, and aquarium owners. Aside from these manageable issues, everything was expected to go smoothly. Yet when CAP water began to course through Tucson's water mains, it wasn't long before all hell broke loose.[20]

The problem was the city's old iron and steel pipes. People first complained about red, brown, or yellow water coming out of their taps. This problem was soon eclipsed by a virtual epidemic of broken pipes and damaged water heaters and evaporative coolers. It turned out that CAP water was highly corrosive to the city's old pipes. (A similar switch in water sources caused lead contamination in Flint, Michigan, two decades later.) The city utility began tinkering with the chemistry of the water, but the damage to homes continued. Public exasperation soon turned into outrage. All CAP water deliveries were halted in 1994 to allow for repairs to the system. The city council subsequently voted not to resume deliveries. In a referendum the next year, voters staunchly outlawed CAP water unless it was treated to the quality of the local groundwater *and* was free of disinfection byproducts from chlorination. A very tall order.[21]

Treating CAP water to these standards would be prohibitively expensive. Fortunately, there was another solution. In the 1960s and 1970s, the city had purchased considerable land west of the city that turned out to have ideal geology for constructing huge recharge basins. It took time to construct the basins, but the system worked. Since then, all CAP water delivered to Tucson is first recharged through these and other basins, where it mixes with native groundwater and, by definition, becomes local groundwater. The system is expensive—just one recharge basin can cost over $1 million—but it's a lot cheaper than the alternative. There's also the added benefit of having built-in water storage, should there be a temporary shutdown of the CAP system.[22]

In spite of these problems, the Central Arizona Project has brought greater water security to this desert state—making the possibility of losing it all the more dire. Everything depends on the river cooperating. The Colorado River is a lifeline to tens of millions of people in the United States and Mexico, as well as meeting major agricultural, mining, and habitat needs. With so much hanging on this wily river, the possibility of a serious and prolonged drought is a frightening thought. Making matters worse, the framers of the Colorado River Compact overestimated the average annual flow of the river because it was based on an unusually wet period. Consequently, the river is overallocated even during a year of "normal" weather.

The current situation is even more worrisome. Beginning in 2000, the Colorado River Basin has been gripped by a historic drought. According to some estimates, this is the most severe event of its kind since the last mega-drought over eight hundred years ago. Some scientists are predicting that droughts of this intensity are the new norm, translating into a "drydown" of western North America.[23]

In the summer of 2015, with the Colorado River Basin in its fifteenth year of drought, Lake Mead fell to its lowest level ever. A declaration of a shortage on the Colorado River appeared imminent, until late-season rain in the Upper Basin saved the day. Managers of the Lower Basin's water supply continue to prepare for the worst.

Las Vegas, which receives 90 percent of its drinking water from Lake Mead, began looking at the possibility of its intake pipes running dry. In response, the Southern Nevada Water Authority allocated more than $800 million to build a three-mile (five-kilometer) long tunnel (through solid rock) to install a new intake pipe at the deepest part of the reservoir. This "third straw" came on line in 2015.

Long before the drought hit, Arizona officials recognized their state's vulnerability. In 1996, the Arizona Water Banking Authority was created to "bank" excess CAP water for the state. As the name suggests, water banking works like a checking account, with water deposited ("recharged") underground to be withdrawn later in time of need. The authority works in one of two ways—it directly recharges excess CAP water, or sells it to farmers at heavily subsidized prices in lieu of the groundwater they would have pumped. By 2014, Arizona had one of the nation's largest water banking programs, with deposits totaling about 1,100 billion gallons (4,200 billion liters) of water for future use. Phoenix, Tucson, and other cities have been developing their own water banking. By 2015, cities in the Phoenix metropolitan area had stored enough water to meet their collective needs for more than two years.[24]

In 2001, the Arizona Water Banking Authority agreed to bank excess CAP water for Nevada. Nevada pays the full price of water delivery and underground storage, as well as the costs to recover the water and deliver it to CAP customers in Arizona. In exchange for its banked credits, Nevada is entitled to an equivalent amount of water from Arizona's share of the Colorado River. By 2012, Nevada had accrued a total of about 600,000 acre-feet of credits—twice its annual allotment to the river. This arrangement demonstrates how the basin states can work together to try to get everybody through the worst-case scenario. Pat Mulroy, former general manager of the Southern Nevada Water Authority, summed up the anxiety: "What scares me to death is that we don't know what the worst case looks like . . . I happen to believe that the rate and pace of climate change is coming upon us faster than we have the ability to get our head around."[25]

Arizona's water management programs have made a significant difference to the state's future, yet there remain "holes in the water bucket," as Sharon Megdal, director of the Arizona Water Resources Research Institute, puts it.[26] A gaping hole in the act is that it largely excludes rural areas, where developers don't have to demonstrate an assured water supply in order to sell land, and farmers continue to pump without restriction. In addition, state law does not recognize the connection of surface water and groundwater.

With reservation lands encompassing 28 percent of the state, another hole in Arizona's water planning is the issue of unresolved Native American water rights. Arizona, however, has made more progress in this area than other states. The Arizona Settlements Act of 2004, the largest settlement of Native American water rights in U.S. history, resolved longstanding disputes over claims in the Gila River system that were of particular importance to the Phoenix metropolitan area. Resolving Native American water rights also provides for the possibility of leasing water to cities.[27]

In the view of many, the biggest problem involves the hundred-year assured water supply in the AMAs. The rule created an obstacle for developers and growing communities that didn't have access to CAP or other renewable water. To address this dilemma, in 1993 the

Arizona Legislature established the Central Arizona Groundwater Replenishment District (CAGRD). The CAGRD collects money from developers and water providers in a three-county area encompassing Phoenix and Tucson in order to purchase renewable water, which it then recharges to cover members' excess groundwater withdrawals. The problem is, there's no requirement to recharge near where the water is withdrawn. Pumping and recharge must take place in the same AMA and be consistent with the AMA goals, but that's all. The law allows developers and water providers to mine groundwater locally, hand a check to CAGRD, and tell it to go buy some water and recharge it somewhere in the AMA. Most of these recharge basins are far from any development. With water becoming harder to find and more expensive, the CAGRD has accumulated a lot of recharge I.O.U.'s. As of December 2014, it had an obligation to replenish (for a hundred years) the groundwater pumped for more than a quarter million homes. Homeowners are responsible for paying these future costs.[28]

Kathleen Ferris is currently serving as executive director of the Arizona Municipal Water Users Association (AMWUA). The members are nine cities and one town in the Phoenix metropolitan area—including Phoenix, Scottsdale, Tempe, and Mesa—that provide water to more than half the state's population. The AMWUA's mission is to foster responsible water stewardship that safeguards water supplies for future generations through conservation and water reuse. In 1999, the AMWUA partner cities, along with others, developed a multimedia campaign called Water—Use It Wisely, which is now the largest water conservation awareness campaign in North America. AMWUA cities also have become leaders in reusing wastewater, now reclaiming 100 percent of the wastewater produced and putting it to beneficial uses—such as for recharging aquifers, irrigating turf, and cooling a nuclear power plant.[29]

Despite Arizona's dramatic growth, the numbers say it all. From 1957 to 2010, Arizona's population increased by 470 percent, while the state's total water use remained the same.[30] By being proactive, Arizona has averted a major water crisis, but it cannot afford to rest on its laurels. Actions taken decades ago have achieved today's water security, but projections indicate that a serious gap will build between supply and demand unless further actions are taken.

Barely a mile from Tucson's interstate is another world, a place of ponds teeming with cattails and bulrushes where ducks float across the still water and great blue heron fish on the banks for frogs and minnows. Mist rises from the water at dawn, creating a surreal tranquility. The rising sun illuminates the nearby Catalina Mountains in rich hues of purple. Songbirds swoop and perch on the water plants and thick brush, chirping, trilling, and warbling in a new day. Dappled sunlight filters through the surrounding riparian woodland of cottonwood and mesquite. Hawks and other birds of prey lift off and soar into the cobalt blue sky. Back out on the interstate, morning rush hour is in full swing—but here in the Sweetwater Wetlands, a world of undisturbed nature and silence prevails.

Tucson's Sweetwater Wetlands was built to recreate part of the vast wetland and riparian woodland habitat that once flourished along the now mostly dried up Santa Cruz River. The wetlands encompass eighteen acres (seven hectares) of ponds and woodland, with trails, observation decks, and a gazebo by the largest pond. This desert oasis teems with life. Since opening in 1998, the Sweetwater Wetlands has attracted nearly two hundred species of birds, mammals, and reptiles—and more than forty thousand humans. Free to the public and open daily from dawn to dark, this nature park is the perfect place for a morning walk or jog, for school fieldtrips and picnics, and for bird-watching.

There's an interesting story of how the Wetlands came to be. In the mid-1990s, the Tucson water utility was cited for failing to meet certain drinking-water requirements for surface-water sources. They had a good excuse. The utility had always used groundwater for its water supply, which didn't require much in the way of treatment, and it was still transitioning to CAP water. But excuses weren't allowed. The Arizona Department of Environmental Quality fined the utility $400,000. Instead of paying the fine, the utility proposed using the money to create a wetland that would filter reclaimed water. The plan was approved. By the time the project was completed, the city of Tucson had spent $1.7 million.

The system was ingeniously designed to use the wetland's natural biological processes to further clean the water after secondary treatment. The water is then released into four recharge basins where it filters into the ground and recharges the aquifer. This recharged water provides seasonal storage so that Tucson Water can meet summertime peak demands for irrigating parks, golf courses, and schoolyards in the metropolitan area.

For most people, the Sweetwater Wetlands is simply a great place to relax and enjoy nature. Just another part of living in this glorious state. But for those in the know, this highly technical and practical facility is saving the city large amounts of water. The Sweetwater Wetlands is a microcosm of Arizona's water story. There's a whole lot more to it than meets the eye.

4

The World's Poorest People

When we restore the water, we restore the human spirit.
—Brian Richter

Lying south of the world's biggest desert (not counting Antarctica and the Arctic), sub-Saharan Africa is the world's poorest and least developed region. In this context "region" is an understatement. The sub-Sahara is so huge that all of China, India, and the United States would fit inside its borders. Yet the size of this region is basically the only thing big about it. By almost any measure, sub-Saharan Africa lacks even the most basic infrastructure to meet the needs of the over 970 million people living here.[1] Roads, electricity, water and sanitation, health services, schools—you name it, they haven't got much of it. Or much that actually works.

At night, from space, Africa lies in near darkness. All of the vast landmass of the sub-Sahara generates only as much electricity as Spain.[2] A few splashes of light prick through the darkness, but they're mostly north of the Sahara along the Mediterranean coast. In the sub-Sahara, where around 85 percent of Africans live, life is mainly rural and poor. But this is a different kind of poverty, requiring an entirely different lens to view it. This is barely-surviving poverty, where many don't. In the sub-Sahara, the average life expectancy is fifty-seven.[3]

The burden is heaviest on women, who spend their lives growing food, collecting wood for cooking it, and hauling water. Only around 16 percent of sub-Saharans have potable water (fit for drinking) piped into their home. Many households must go out and find it, then haul it home.[4]

This "water walk" is global. Every day around the world, women and girls spend an estimated 125 million hours finding and carrying water.[5] But the sub-Sahara is the world's water-walk hub. Many African women and their daughters spend much of their day walking miles to a water source, then walking back over primitive rocky roads, often in worn-down flip flops, with more than forty pounds (eighteen kilograms) of water on top of their heads. The girls carry lesser amounts, but compared to body size the loads are relative.

Hauling heavy loads of water on top of your head over long distances every day takes a toll on the body. Many women develop painful muscular and skeletal problems. Sub-Saharan Africa also has the world's highest birth rate. This means that during those numerous pregnancies, and until the child is strong enough to go the distance without being carried, the water walk is a daily marathon of torture. Men and boys seldom pitch in and help. Fetching water is women's work.

This gets worse. While millions of girls have to drop out of school to help their mothers carry water, and soon grow into women who lose their lives and health to the water walk, much of that water is so filthy that we water-pampered Westerners wouldn't let our dogs drink it. But they don't have a choice. Survival doesn't get any more basic than water. When that's all there is, you drink it. And your children drink it. They may die from drinking it, but they definitely will if they don't.

Drinking filthy water results in untold disease—typhus, cholera, trachoma (which can lead to blindness), and the big scourge, diarrhea. The world's second biggest cause of death among children under five is diarrhea. At any given time, roughly half of all the world's hospital beds are occupied by patients suffering from water-related diseases. The numbers are much higher in the sub-Sahara, with the crucial difference being that hospitals, and hospital beds, are about as scarce as clean water.[6]

Many of these water-related illnesses result from a lack of sanitation. Again, this problem is most severe in sub-Saharan Africa, where

Dahery, Tiana, and their friends at school in Madagascar. Clean water for drinking and washing hands is critical to the lives and well-being of these children. *Source:* WaterAid/Ernest Randriarimalala. Used with permission.

over 60 percent of the population doesn't have even basic sanitation, and about a quarter of the population practices open defecation.[7] The result is a virtually unbroken cycle of walking to get water, and making yourself and others sick in the process.

All this water-related illness takes a heavy toll on education. About 30 percent of the world's schools don't have an adequate water source and one-third lack adequate sanitation. As these things are defined, "adequate sanitation" is a step or two above basic sanitation (an outhouse, or some version thereof)—yet it is still a far cry from the private, pristine, porcelain toilets and sinks that Westerners take for granted. For girls who haven't already dropped out to help their mothers carry water, the onset of puberty usually means the end of their education. Inadequate sanitation means no bathroom.[8]

Economists have coined a name for this tragic interconnected problem. They call it the "water-poverty trap," meaning that without free access to clean water, the chances of breaking out of poverty are slim to nonexistent. Globally, one out of every ten people lack access to an improved water source. ("Improved" doesn't necessarily mean safe.) Many poor people who lack access and don't collect dirty water from ponds and rivers have no choice but to buy water (of questionable quality) from street vendors or tanker trucks at inflated prices. The poorest of the poor may spend more than half their daily wages on water.[9]

In 1948, the newly formed United Nations issued its Declaration of Human Rights, which stated that "Everyone has the right to a standard of living adequate for the health and well-being of himself and of his family, including food, clothing, housing, and medical care."[10] The United Nations was created as part of the Marshall Plan to get Europe, especially Germany, back on its feet so there wouldn't be another War to End All Wars.

A decade later, with the United Nations' original mandate winding down in postwar Europe, the U.N. General Assembly proclaimed the 1960s to be the First U.N. Development Decade. The new mandate was now global and ignited a flurry of development summits. Unfortunately, processes for monitoring development targets and producing coordinated plans of action somehow got overlooked. The Development Decade fizzled out with little to show for it.[11]

In 1990, summitry returned with a bang. There was the World Conference on Education for All, the U.N. World Summit for Children, the U.N. Conference on the Least Developed Countries, and a Conference on Drug Problems. In 1992, the U.N. Conference on Environment and Development, held in Rio de Janeiro and also known as the Earth Summit, was a massive effort to draw the world's attention to the connection between development and the environment. The world took note, but conference leaders failed to achieve a consensus on such major issues as climate change and deforestation. That same year, the International Conference on Nutrition resulted in a unanimous commitment to "freedom from hunger." There was little follow-up. In 1993, Vienna hosted the World Conference on Human Rights. No word on that one. The 1994 International Conference on Population and Development broke down into heated debates over women's empowerment and right to abortion. In 1995, Copenhagen's World Summit for Social Development was subsequently criticized for focusing on symptoms instead of underlying causes. That same year, Istanbul hosted the Second U.N. Conference on Human Settlements. In 1996, Rome's World Food Summit was followed by the United Nations' announcement that the entire year was now an International Year for the Eradication of Poverty—soon to be expanded into the International Decade for the Eradication of Poverty.[12]

Despite some progress, there was a sense of summit fatigue. There were just too many recommendations on too many subjects, with too many national leaders making grand promises. There was also a big problem. Aid agencies around the world were in a long-term decline, limping along on about half the funding they had in 1960. The Cold War was over, and there was no longer a perceived need to use foreign aid to buy allies in resource-rich poor countries.

The United Nations' reputation also was in decline, and on the line, due to all those summits and not much to show for it. U.N. Secretary-General Kofi Annan was determined to make the upcoming Millennium Assembly a stunning success and bring the world together in a reenergized frame of mind to tackle global problems. The United Nations also needed to demonstrate its value to the most powerful members—the United States and other permanent members of the Security Council—so it would continue to be financed.

In the spring of 2000, Annan delivered his "We the Peoples" speech to the General Assembly, declaring that "we must spare no effort to free our fellow men and women from abject and dehumanizing poverty."[13] At the Millennium Summit in September, the largest gathering of world leaders in history adopted the U.N. Millennium Declaration, committing to slash extreme poverty in half by the year 2015.

Reaching consensus on the declaration's goals involved a long, politically fraught tug-of-war among donor countries. The eventual outcome was eight goals: eradication of extreme poverty and hunger; universal primary education; gender equality; a decrease in child mortality; improved maternal health; prevention of the spread of HIV/AIDS, malaria, and other diseases; environmental sustainability; and (the only goal that didn't have a 2015 deadline) developing a global partnership for development.

The water-poverty trap directly affects nearly all these goals, but it didn't become a Millennium Development Goal (MDG). Instead, water was relegated to two targets under the environmental goal: to halve, by 2015, the proportion of people without sustainable access to safe drinking water and to basic sanitation.[14]

Catarina de Albuquerque, the first U.N. special rapporteur on the right to safe drinking water and sanitation, was tasked to go on three fact-finding missions each year. "What I often see in developing countries," she reported, "is that donors often prioritize financial support to huge systems that they cannot maintain [and] that are not helpful to rural areas. Additionally, the MDG water and sanitation targets don't specify whom to target, so it can be better-off people."[15] This problem is well documented among aid agencies, development organizations, and water experts—the hardest to reach, the rural and desperately poor, are left out time and again.

Albuquerque raised another issue: the wording of the drinking-water target had been changed, so that "safe" became "improved." "When it comes to quality this is obviously a crucial criterion, which is not currently in the proxy indicators for the MDG targets," she explained.[16] The lack of a reliable water-quality test that can be performed in remote areas of developing countries makes it difficult to test the water. The thinking was that an improved source is "likely" to provide safe drinking water, which is sometimes true. Nonetheless, an improved water source isn't necessarily a safe one.

What this word-smithing did mean was that the United Nations and the World Health Organization (WHO) were able to celebrate the early completion of the drinking-water target. In the spring of 2012, they announced that an additional two billion people had gained access to improved drinking water, while "only" 783 million people around the globe still relied on unimproved water sources such as streams, ditches, or unprotected wells.[17]

Shortly before this announcement, a study out of the University of North Carolina, Chapel Hill revealed that many sources of "improved" water failed a safety test conducted in five countries (two in sub-Saharan Africa) spread over three continents. Extrapolating these findings to the rest of the globe is difficult, but their estimate gave a ballpark figure of 1.8 billion people (28 percent of the world's population) still drinking unsafe water. This is far in excess of the 783 million (11 percent) that the United Nations and World Health Organization would announce.[18]

The World Bank has made considerable investments in rural water supplies. Between 1978 and 2003, the bank lent approximately U.S. $1.5 billion to this sector. Wells were dug or drilled and fitted with hand pumps, and piped water systems were constructed. Nonetheless, rural water-supply coverage lags significantly behind urban supplies. In sub-Saharan Africa, the disparity was much greater, and getting worse. The number of rural dwellers without access to an improved water supply *increased* by almost thirty million people from 1990 to 2006.[19]

There's plenty of blame to pass around. The World Bank, many governments, donors, and NGOs have tended to focus on numerical targets (so many wells drilled this month) instead of paying attention to what's happening on the ground. And much of what *is* happening is the same old story of corrupt governments siphoning off Millennium Development Goal money to support the ruling elite.

But that's only part of the problem. Due to a lack of oversight, standards, and regulation, the sub-Saharan countryside has been turned into a virtual Wild West of drilling. In general, anything goes. How this plays out is that almost anyone (whether they're qualified or not) can decide to "do good," turn up in a village and improve the water supply as they see fit (whether they know what they're doing or not), and work according to their own standards (if they have them) with little or no input from the people they are trying to help.[20]

Too often a donor team goes to a village, drills a well, and goes home. Many of the wells never work properly, or soon stop working. The donors typically neglect working with the villagers to decide who will oversee maintenance and train them how to do it. The consequences of all this laissez-faire drilling are often serious. What was supposed to be a well now becomes just a hole in the ground, or worse, a dumping place where contaminants take a fast track to groundwater. What could have been a village's source of safe drinking water becomes a contaminated aquifer. And so it's back to the water walk for the women and girls.

In spite of the problems, the Millennium Development Goals fostered a global agenda to improve the lives of disadvantaged people. In 2015, the United Nations initiated a follow-on set of Sustainable Development Goals for the next fifteen years. The sixth goal is to "ensure availability and sustainable management of water and sanitation for all," including a target of equitable access to safe and affordable drinking water.[21] The need to support and strengthen local community participation is recognized as essential to achieve this goal.

Against this backdrop are groups and individuals who recognize the need to work with responsible professionals, develop a partnership with the community they're trying to help, and make sure someone nearby is trained in maintenance and repair so the well will continue to work after the donor team has left. The goal is to leave behind capacity, not just holes in the ground.

Among these groups is the Rural Water Supply Network (RWSN), a global network of more than six thousand professionals working to raise standards in rural areas of developing countries. The RWSN has made considerable strides in fostering technical and professional competence through development of best practices and direct assistance. They also lead the way in innovative methods of fostering communication. Facilitated email discussion groups allow those working on rural water supplies to share experiences and expertise on any matter related to siting, drilling, or maintaining wells. The RWSN also hosts webinars where people working in sub-Saharan Africa and other developing regions share their experiences on practical topics such as rainwater harvesting and well construction.[22]

Many individuals are also making a difference. A good example is Rochelle Holm, an environmental scientist from Richland, Washington, who has been working in Malawi since 2007.[23] Located at the southern end of the Great Rift Valley, Malawi is one of the poorest countries in sub-Saharan Africa. Less than 1 percent of the population has access to a university education—one of the lowest rates in the world. Those in the rural villages are trying to eke out a living through subsistence farming. The government's official national water policy is "Water and sanitation for all, always." While improvements have been made in providing access to improved sources of water, many people in rural areas still rely on polluted streams and turbid ponds, or an improperly sealed well, for their water supply. In addition, the villages are surrounded by shallow latrines, sometimes hundreds of them, which serve as point sources that are further contaminating the water supply.

Through the missionary work of her church, Rochelle Holm obtained a grant from a research and education foundation operated by the National Ground Water Association (NGWA) to help one of these villages obtain a safe water supply. The NGWA is the world's largest organization of groundwater professionals, and is committed to providing guidance for the responsible development and management of groundwater resources. With her grant, Holm chose the village of Geisha in northern Malawi, "because they had demonstrated the crucial first step." In other words, this village had gone beyond talking about a clean water supply and had taken the initial steps to make it happen. They also had the necessary leadership and commitment to the project that would help ensure success.

After assembling an experienced technical team, Holm began to build a partnership with the community. Reverend Levi Nyondo, the village project leader, formed a committee of both men and women to organize food and housing for the drilling team and to oversee obtaining the raw materials (three tons of sand, one ton of gravel, and 2,500 clay bricks). Five people traveled from the United States to Geisha for the fieldwork, and were supported by a stateside team providing ongoing technical input for what looked like the best places to site the well. But here is where Holm draws the line. "The final decision of where to place the well should *never* be the donor's role," she emphasizes. "If the best yielding Malawian well is not used due to geographic, political, or religious considerations, the project fails." The village contracted with "Lucky" Penumlungu, a water-well consultant with thirty-five years of drilling and consulting experience and well-established contacts with the government. Finally, the drilling commenced and the village soon had a well that was providing a plentiful supply of clean water.

To assure long-term success, the well committee was trained in maintenance. By including women on the committee, conscientious maintenance was virtually guaranteed. The committee also received training in proper sanitation practices, such as not allowing a latrine near the well. Nyondo continued as project leader, and would direct any future questions or concerns to Penumlungu.

Rochelle Holm now lives in Malawi, where she manages the Centre of Excellence in Water and Sanitation at Mzuzu University. She continues doing what she considers her calling—helping people learn how to develop and maintain their own water projects.

There's also the extraordinary story of Ryan Hreljac, who grew up in Ottawa, Canada.[24] When Ryan was in first grade, his teacher gave a lesson on poverty. She explained that people in the world were dying because they didn't have clean water, and that many people in Africa have to walk for hours, sometimes just to get dirty water. "Before that day in school," Ryan explains, "I figured everyone lived like I did and so when I found this out, I decided I had to do something about it." He started doing extra chores to raise money to help, and telling everyone he knew about the water crisis. A year later, when he was in second grade, Ryan had raised enough money to build a well at a school in Uganda. When he was nine years old, he established Ryan's Well

Foundation—a Canadian-registered charity focused on providing clean and safe water in developing countries.

Since that first well was constructed in Uganda, Ryan's Well Foundation has helped build more than 875 water projects and 1,120 latrines, bringing safe water and improved sanitation to more than 850,000 people. For every project, a local water committee is organized that trains members of the community how to manage and maintain the well over the long term. The foundation then provides ongoing monitoring and evaluation of the project.

One of its most recent projects is the Adwir Health Clinic in northern Uganda. The clinic serves a wide range of medical needs, including delivery of babies, but it lacked both water and toilets. Ryan's Well Foundation chose this clinic as its Seasonal Giving project and began raising money to build a well, safe latrines, a washing room, and handwashing facilities. With water and sanitation in place, the clinic is now able to attract a staff of doctors and nurses.

In addition to working for his foundation, Ryan Hreljac travels around the world speaking on water issues and the importance of making a difference. He tells his audiences, "We are all ordinary people if you think about it, but when you have a passion and you invite other ordinary people to join in, you become a community. And it's with a community of ordinary people that you can make extraordinary things happen." All very true, except for the fact that Hreljac, now in his mid-twenties, is far from an ordinary person.

Many other individuals and groups are working to build capacity in developing countries. Stephen J. Schneider, a drilling contractor from Oregon, led a concerted effort to prepare "Water Supply Well Guidelines for Use in Developing Countries," a manual that has been translated into multiple languages, including Swahili. Also deserving of special mention are groups such as Hydrogeologists without Borders (HWB) and the Ann Campana Judge Foundation, which have taken "hydrophilanthropy" and water "capacity building" to Central America. The Ann Campana Judge Foundation honors Ann Campana Judge, who died aboard American Airlines Flight 77 when it was crashed into the Pentagon on September 11, 2001. She had traveled to many developing countries as part of her work for the National Geographic Society and was a great supporter of students and those less fortunate than herself.[25]

Paul Polak is considered the father of market-based solutions to poverty. For over thirty years, he has helped millions of the world's poorest people lift themselves out of poverty. Polak's philosophy for sustainable poverty eradication is pragmatic and to the point: we cannot donate people out of poverty. Polak explains that viewing the poor as victims in need of handouts just breeds dependency. "To move out of poverty, poor people have to invest their own time and money. Real poverty eradication comes by treating the poor as workers and customers," he believes.[26]

This view is now mainstream thinking in the world of international development, but during the first twenty years of Polak's work, development leaders were extremely resistant to his belief that you can, and should, sell things to poor people at a fair market price instead of giving them things for nothing. For over three decades, Polak has worked with talented and innovative product designers who have developed affordable, high-quality tools for the world's poorest people by applying his favorite mantra—the ruthless pursuit of affordability.

Polak's 2013 book, *The Business Solution to Poverty,* explains the design principles for how to supply the very poor, ethically and effectively, with vital necessities at a fraction of the usual cost. Polak argues that there is a misplaced perception that the marketplace serving the 2.7 billion bottom-of-the-pyramid customers requires products that work poorly, break easily, and look cheap. "Nothing could be further from the truth," he emphasizes. "Products that are attractive to poor customers must indeed be affordable, but they also need to work well and look good."

"Not that any of that is simple!" Polak is quick to add. "Product design for the bottom billions requires a new mind-set." He calls it "zero-based design," which means abandoning your assumptions, starting from scratch, and learning what's really necessary and feasible. "Then think big and act big."

In 1981, Polak started International Development Enterprises (IDE), a nonprofit dedicated to ending poverty in the developing world by helping farm families access the tools and knowledge they need to increase their income. "Most problems are complex," Polak explains. "If you want to understand a complex problem, you have to reach a thorough understanding of each of its root causes and how they interact. But finding a practical solution requires a different strategy. It's more a matter of finding the simplest *lever* capable of producing the biggest positive result."

Not surprisingly, the first step out of poverty for most of the world's poorest farmers is to find an affordable way to bring water to their crops. Since groundwater is often the most easily accessible source of water for irrigation, they need access to an affordable means of lifting it to the surface. For Polak, the first lever out of poverty was a simple suction-treadle pump. The Norwegian engineer Gunnar Barnes designed a pump that could be activated by manual labor, something like pedaling a bicycle. It was easy to operate and maintain, and cost less than a sack of rice. IDE marketed it through the local private sector.

The biggest challenge was to persuade tradition-bound farmers with no spare pennies to take a risk. To combat this resistance, Polak and a growing staff developed a marketing program that was designed for illiterate and very poor people with no access to the media. The advertising included wall posters, troubadours performing at village markets, and rickshaw processions. The big hit was a free ninety-minute film that featured a poor farmer who buys a treadle pump and gets the girl. The film was soon playing to an audience of a million people a year.[27]

In conjunction with the marketing program, IDE developed a network of manufacturers, village dealers, and well drillers who had graduated with a certificate from an IDE training program. The treadle pumps were sold at an unsubsidized market price of $25—compared to the $500 cost of the cheapest diesel-powered pump and well assembly available at the time. By 2008, IDE had sold more than two million

Cambodian woman treadling. Reprinted with permission of the publisher. From *Out of Poverty*, copyright © 2008 by Paul Polak, Berrett-Koehler Publishers, Inc., San Francisco, CA. All rights reserved. www.bkconnection.com.

treadle pumps to poor farm families, increasing their net annual income by more than two hundred million dollars.

Polak tells the story of one of the thousands of poor farmers he has talked to over the years. "I met John Mbingwe in an isolated rural village in Zambia in 2001. For years, he had been growing vegetables by carrying water in buckets from a crude five-foot-deep well he and his wife had dug. But carrying water by bucket is very hard work, and he and his wife worked long hours to irrigate an eighth of an acre of vegetables. Then they borrowed enough money to install a treadle pump and, with less labor than it took to water an eighth of an acre by bucket, he and his family suddenly found they could produce a full acre of vegetables. Within a year they had paid off the loan for the pump, increased their net annual income from three hundred to six hundred dollars, and were on their way to earning more."

Once the treadle pump was successfully launched, the next lever was designing a drip irrigation system for very poor farmers. It took five years of trial and error before this simple idea could be transformed into a low-cost system that worked. The final product met all of Polak's criteria. Like a LEGO set, it's infinitely expandable. As poor farmers increase their income and acreage under cultivation, they can easily add on to the drip system. It's also simple to maintain—the holes can be easily unplugged with a standard safety pin. And it passed, with flying colors, the ruthless-pursuit-of-affordability requirement. The starter kit costs three to five dollars. IDE's drip irrigation system was soon in widespread use. "Early market demand," Polak explains, "suggests that the global demand for low-cost drip systems will reach at least 10 million families, increasing their net annual income by $2 billion a year."

With the treadle pump and drip irrigation system now providing affordable and reliable access to groundwater, the next lever was crop selection. IDE set up a program to help farmers pick four or five high-value, off-season fruits and vegetables that would grow well in their area. Then they developed private-sector supply chains to sell the seeds and the fertilizer that farmers would need to grow these crops.

Closing the loop, the final lever was the development of connections to larger markets. "Unfortunately, the markets where dollar-a-day people are buyers and sellers have more holes than a barrel of Swiss cheese," Polak says. "Most markets are so far away they can't be reached, or don't exist at all. How can they increase their income if they don't have access to markets where they can sell what they produce at a profit?"

In response to this need, IDE helped poor farmers organize and pool their crop yields, enabling them to sell to larger food companies and organizations. They brokered deals with farmer's markets and supermarkets, and set up collection points where vegetable traders could buy the produce in bulk amounts. By unleashing the entrepreneurial spirit of the world's poorest farmers, IDE is helping them to become competitive commercial farmers.

Today, IDE has more than 580 employees working in eleven countries on three continents. Paul Polak will tell you that's just the beginning: "It's clear that without a revolution in thinking and practice on the part of the development community, the business community, and poor people themselves, we will never be able to end poverty. But if we learn to listen to poor people, understand the specific contexts in which they live and operate, and find ways to harness their entrepreneurial energy to increase their income, I have no doubt that at least 500 million families now surviving on about a dollar a day will find practical ways to end their poverty within one generation."

It just takes a lever. And groundwater.

5

Not All Aquifers Are Created Equal

To be absolutely certain about something,
one must know everything or nothing about it.
—Olin Miller

How much groundwater is there? The numbers are impressive. More than 95 percent of the world's unfrozen freshwater (not tied up in glaciers and polar ice) is groundwater. The United States has groundwater reserves of at least 33,000 trillion gallons (125 trillion cubic meters). This is approximately as much water as the Mississippi River has discharged into the Gulf of Mexico during the past two hundred years.[1] These statistics make great conversation among friends, but unfortunately they're misleading. Much of the Earth's groundwater is inaccessible, cannot be developed without causing problems, or is too contaminated for human use.

The Great Lakes provide a good analogy. These massive water bodies are a remnant of continental glaciation from the last ice age, and contain about 20 percent of the world's fresh surface water. Yet only a tiny percentage of this vast water resource can be withdrawn without causing serious problems for navigation, docks, and owners of waterfront property. As a result, the Great Lakes states and bordering Canadian provinces aggressively guard this resource from being diverted outside the basin, no matter how "small" the proposed diversion.

The same problem exists with groundwater. Removing even a small percentage of the total volume of groundwater in an area often sets into play any number of possible consequences. The most obvious problem is that pumping costs rise as groundwater levels decline and eventually may become cost prohibitive. Yet there are many aquifers where other factors come into play before economics. Groundwater pumping in coastal areas can be limited by seawater intrusion—pump too much and wells produce water too salty for use. Then there's land subsidence. Houston sits on top of a vast groundwater resource, but the subsidence caused by groundwater pumping has markedly increased the city's vulnerability to flooding and tidal surges. As a result, Houston converted its water-supply system from groundwater to surface water at considerable expense. A common limitation on the use of groundwater arises from its connection to surface water. Even a small change in groundwater storage can have significant impacts on streams, springs, and wetlands. (In spite of this interdependence, most U.S. states have separate laws governing groundwater and surface water.) Thus, at one and the same time, groundwater is a vast resource and a limited one. The challenge is to recognize both characteristics and act accordingly.[2]

Aquifers are best understood on a case by case basis. Aquifers and aquifer systems (a sequence of geologic units that behaves as a single hydrologic system) come in many shapes and sizes. They underlie areas ranging from a few square miles to tens of thousands of square miles. Thicknesses range from tens of feet to several thousand feet. Aquifers and aquifer systems also differ greatly in their ability to store and transmit groundwater. Geology and climate (present-day and, in many cases, far in the past) largely determine the groundwater situation. Let's look at some examples.

In broadest terms, aquifers are classified as either unconfined or confined. An unconfined aquifer has the water table as its upper boundary and is in contact with the atmosphere through an unsaturated zone. In contrast, a confined aquifer is overlain by a geologic formation with low permeability—often clay or shale.

An aquifer's response to pumping is different depending on whether it is unconfined or confined. In an unconfined aquifer, the principal source of water is dewatering of the aquifer by gravity drainage of the pores. In contrast, water pumped from confined aquifers is derived from aquifer compression and water expansion as the hydraulic pressure is reduced. Pumping the same quantity of water from confined aquifers results in larger water-level declines covering much larger areas when compared to unconfined aquifers. An additional complication arises because the drawdowns in a confined aquifer will induce leakage from adjacent confining units.

When a well is drilled into a confined aquifer, the water level in the well rises above the base of the confining unit. If enough pressure exists, the water level will rise above the land surface, and the well is called artesian.[3] These flowing wells can deliver spectacular shows.

The term artesian comes from the French region of Artois, where some of the first artesian wells were drilled. Some, such as the famous artesian well at Grenelle, Paris, were major engineering feats. Geological studies had suggested that sands outcropping to the east of Paris also could be found at depth under the city. As it turned out, obtaining water from these deep sands required marathon tenacity. Drilling continued for more than eight years and to a depth of 1,800 feet (548 meters) before hitting the jackpot in 1841—when water began shooting more than a hundred feet (thirty meters) into the air. Controlling this geyser required encasing it in a thirteen-story iron tower that became a major tourist attraction, perhaps even serving as some inspiration for the Eiffel Tower (the world's most heavily visited paid attraction).[4]

Artesian wells were once common around the world. In the United States, numerous artesian wells were drilled along the Atlantic coast, as well as in California, Texas, and the Dakotas. Along the coast of Los Angeles, the artesian pressure was sufficiently high that sailors could dip buckets and collect freshwater without having to go ashore. In 1876, a successful artesian well in the Fort Worth area of Texas made worldwide news, setting off a drilling craze. After hitting the "Dallas Geyser," which produced a million gallons a day, nearby Dallas made plans to drill 270 wells to pump water to the Trinity River and transform this inland city into a seaport. With wells left uncapped and water flowing down the streets, the boom was short-lived. By 1894, only a handful of wells in the area were still flowing.[5]

The world's artesian wells had their heyday in the late 1800s and early 1900s. Many essentially bled to death as they went uncapped. In many places where artesian wells once gushed, water levels are now hundreds of feet below the land surface. The Great Artesian Basin in Australia is a rare example where efforts have been made to reverse this course.

The Great Artesian Basin underlies about one-fifth of the Australian land mass. This multilayered aquifer system is comprised of sandstone aquifers interbedded with shale confining beds. The aquifer system is up to ten thousand feet (three thousand meters) thick. Water percolates slowly through the aquifer, and in places has been dated at more than a million years old.[6]

The Great Artesian Basin is a groundwater analogy to the Great Lakes, holding enough water to cover the entire Earth's land mass under 1.5 feet (0.5 meter) of water. There's possibly enough groundwater in the basin to meet Australia's needs for 1,500 years. The problem is that less than 1 percent of this water is considered exploitable without seriously disrupting springs, and without water levels falling to excessive depths for affordable pumping. Springs provide a natural outlet, particularly along the southern and western edges of the basin. For thousands of years, Australia's indigenous people have relied on these springs as their source of water. The springs also sustain many unique flora and fauna.[7]

The Great Artesian Basin underlies arid and semi-arid terrain where surface-water resources are not only few and far between, but also unreliable. The basin's wells provide the only significant source of water for towns, mining, and sheep grazing. The groundwater is generally unsuitable for irrigation because of its high sodium content.

A drilling boom in the Great Artesian Basin began in the late 1800s. Many of the wells diverted water into miles of earthen drains, opening up extensive tracts of new grazing country. More than 95 percent of this water was lost to evaporation and seepage. As a result of falling water pressure, more than a thousand natural springs dried up and at least a third of the artesian boreholes ceased flowing.[8]

In 1989, Australia began an effort to reverse this situation, as water was needed for growth in the mining, petroleum, and geothermal industries. By 2013, more than 1,100 boreholes had been capped and nearly 15,000 miles (25,000 kilometers) of open earthen drains were replaced with pipe. The water savings are equivalent to the residential use of about two million people a year.[9]

In 2012, a team of British scientists presented the first continent-wide, quantitative maps of Africa's groundwater storage and potential well yields. Their findings spread through the news media, which proclaimed that the African continent "is sitting on a vast reservoir" and has a "sea of groundwater reserves," to cite just two headlines.[10]

All this media hoopla was a good thing for getting people to think about the importance of groundwater. As is often the case, however, the media story was quite different from what the scientists were actually saying. News coverage focused on the scientists' overall estimates of the amount of groundwater in Africa, whereas the study team had stressed that Africa's groundwater is unevenly distributed and that the potential for large-scale groundwater development is very limited throughout much of the continent.

Africa probably has the greatest spatial variability in its water resources of any continent on Earth.[11] About 70 percent of Africa's groundwater lies beneath the Sahara Desert—a legacy of past wetter climates and a mostly nonrenewable resource today. In contrast, sub-Saharan Africa has much less groundwater, but most of it is renewable from year to year, with a few exceptions, such as the Kalahari Desert.

Sub-Saharan Africa's geology consists of four basic rock types: Precambrian basement rocks, consolidated sedimentary rocks, unconsolidated sediments, and volcanic rocks.[12] Precambrian basement rocks, comprised of crystalline igneous and metamorphic rocks, cover about 40 percent of the region. As the term basement rocks implies, these are the continent's oldest rocks, having formed more than a half billion years ago. Depending on the extent of weathering and fracturing, basement rocks can yield virtually no water or modest amounts of groundwater. Weathering creates pore space for storing groundwater, and fractures provide the "highways" for transmitting water to wells. If weathering and fracturing are sufficient, basement rock can provide ample water for domestic supply and livestock. With more than two hundred million Africans living on top of basement rocks, finding suitable locations for drilling wells in this terrain has become a major focus.

Consolidated sedimentary rocks (such as sandstone, limestone, and mudstone) comprise about 32 percent of sub-Saharan Africa. Sandstones and limestones are significant sources of water in semi-arid and arid parts of the region, such as the Kalahari Basin. Unfortunately, about two-thirds of the consolidated sedimentary rocks in sub-Saharan Africa are mudstones, which make relatively poor aquifers. Nonetheless, in some places, mudstones can be a useful source for a village's water supply and other local uses.

Unconsolidated sediments make up about 22 percent of the region. These geologically young deposits have not been substantially compacted over time, often resulting in plenty of pore space for groundwater storage and good transmissive properties that allow water to move easily. This type of aquifer underlies many of the world's most productive agricultural regions, including California's Central Valley and the High Plains in the United States, the Indus Basin in Pakistan and India, and the grain belt of North China. In the United States, 86 percent of the groundwater used for irrigation is pumped from unconsolidated (and semi-consolidated) sediments.[13] Sub-Saharan Africa's unconsolidated sediments, however, are largely restricted to small aquifers along streams, the Congo Basin, and the deltas of large rivers, such as the Niger River in West Africa. Another limitation is that many African rivers are laden with fine-grained sediments, rather than coarse sand and gravel, which means that well yields from alluvial aquifers are reduced.

Volcanic rocks make up the final 6 percent. Found mostly in East Africa's Great Rift Valley, these complex aquifer systems are difficult to develop and the water can present major health-related problems due to high concentrations of naturally occurring fluoride. Nonetheless, these are important aquifers because they underlie many of the poorest and most drought-vulnerable areas.

Most poor rural households in sub-Saharan Africa depend on groundwater for domestic use and/or livestock watering. Groundwater irrigation is still relatively limited, but growing in use as a result of improved access to low-cost technology for pumps and drilling services, as well as new market opportunities for produce.[14] The major groundwater problems in sub-Saharan Africa are lack of access and contamination, rather than overdraft. But there's a fine line to walk. Pumping too much for large-scale irrigation could draw down water levels below the depths of the shallow wells that most people depend on for their basic needs. Of rising concern are intercontinental land grabbers (and water grabbers) from China and other countries intent on turning sub-Saharan Africa into their bread basket. Therefore, when headlines such as "Africa has a sea of groundwater reserves" appear in the news, there's reason to be concerned that the wrong message is being delivered.

Most aquifers are partially renewable, albeit over a period of many years. At the extreme end are nonrenewable aquifers, where negligible recharge has occurred in thousands of years. Most nonrenewable aquifers are located in hot, arid regions such as North Africa, the Arabian Peninsula, and the Australian Outback. An exception is Russia's West Siberian Basin, where permafrost provides a barrier to recharge. Development of nonrenewable aquifers is equivalent to the one-shot deal of mining groundwater. This mined water is usually fossil water that accumulated long ago, when the climate was wetter.

Not all aquifer systems containing fossil water are nonrenewable. Some are replenished by recharge from streams or manmade recharge facilities. Others, such as the southern High Plains Aquifer in the United States, are only partly renewed by modern recharge. Over much of the southern High Plains, large-scale pumping has the same end result as if it were a nonrenewable aquifer. It's just taking longer.

Many people advocate for the planned depletion of nonrenewable groundwater reserves as a way to stimulate economic and social development in the near future.[15] The argument is that people could initially rely on irrigated farming for their livelihood, which would allow them to gradually transition to higher-value economic activities that are not water intensive. The downside to this thinking is that such an approach requires careful planning and an appropriate exit strategy. It also overlooks the human tendency to exploit resources today and let future generations deal with the consequences.

A few nonrenewable aquifers have immense groundwater reserves. The Nubian Aquifer System underlies the eastern half of Libya, much of Egypt, and parts of northern Chad and Sudan. Consisting of interconnected sandstone aquifers, this mega-aquifer has a maximum thickness of several kilometers and covers an area four to five times that of the High Plains Aquifer. Groundwater in the Nubian Aquifer System was mostly recharged prior to the Last Glacial Maximum, about twenty thousand years ago. Some was recharged more than a million years ago. Libya and Egypt are today's primary beneficiaries of this ancient endowment.[16]

Libya is one of the driest countries in the world, with only a narrow coastal strip receiving more than four inches (100 mm) of rain a year. With demand far outstripping supply, and its coastal aquifers contaminated with seawater from excessive pumping, Libya turned to expensive desalination plants to meet its water needs. Yet under the desert to the south is a vast groundwater resource exceeding that of any other country in Africa. First revealed during oil exploration in the 1950s, this groundwater is stored in the Nubian Aquifer System in the eastern half of the country and in two massive aquifers in the western half. Virtually all of this groundwater is nonrenewable.

Libya's eccentric former dictator, Muammar Qaddafi, became intent on using these aquifers to provide water for Libya's citizens and to make the country self-sufficient in food production. There was one major impediment—all this groundwater was far from the coast where most of the population lives. Not one to shy away from such a challenge, Qaddafi boasted that he would create the eighth wonder of the world—the Great Man-Made River—to bring water to the coastal cities from more than a thousand wells tapping the desert aquifers.[17]

To avoid the colossal evaporation losses from a surface canal, the Great Man-Made River consists of more than 2,300 miles (4,000 kilometers) of buried concrete pipe. The project officially began in 1984, when Qaddafi laid the foundation stone for a pipe production plant at Brega. Less than a decade later, water flowed from the eastern well fields to Benghazi. Additional phases delivered water to Tripoli and other coastal cities.

With fossil water flowing into Libya's cities, the government turned toward Qaddafi's second goal of reducing the nation's dependence on imported food. Water at highly subsidized prices was provided to farmers to grow wheat, oats, corn, and barley—as well as specialty crops such as almonds and grapes.

The Great Man-Made River project clearly improved people's lives. "For the first time in our history, there was water in the tap for washing, shaving and showering," a senior official in the Great Man-Made River Authority told the BBC.[18] Using nonrenewable groundwater for farming in Libya's brutal desert is another matter. The government claims that the water could last several thousand years. Others estimate depletion of exploitable groundwater reserves within just sixty to one hundred years. Ultimately, the short-term gains may make the country more vulnerable in the future.[19]

The Kingdom of Saudi Arabia illustrates this point. During the 1970s, in an effort to gain food self-sufficiency and improve the livelihoods of rural communities, the Saudis heavily subsidized irrigation for wheat and other crops. By 1992, Saudi Arabia was the world's sixth largest wheat exporter. The problem was that an equivalent amount of wheat could have been bought in the global market at a fifth of the cost of the $2 billion in subsidies *and* prevented massive depletion of the country's groundwater reserves. By the turn of the millennium, Saudi Arabia had racked up about two-thirds of the world's depletion of nonrenewable groundwater. The country has since moved away from its goal of national wheat self-sufficiency and now plans to import all its wheat.[20]

Nonrenewable aquifers, like those in Saudi Arabia and North Africa, are a legacy of past climates and cannot be recharged under any foreseeable climate conditions. Their vulnerability to climate disruption has long since passed. The same cannot be said for other aquifers around the world.

The Edwards Aquifer in south-central Texas is the primary source of drinking water for San Antonio (the nation's seventh largest city) and much of the surrounding region. The aquifer is made up of carbonate rocks that are relatively easily dissolved by water. As water percolates down through cracks in the rocks, it opens up fissures and channels, creating systems of underground drainage. This type of aquifer, known as "karst" after the Kras region in the Slovenian Dinaric Alps, is not rare. About 15 percent of the Earth's land surface is covered by karst, including much of Florida and the Ozarks in the United States, as well as large areas of South China, Europe, and the Caribbean. The Nullarbor (meaning "no trees") Plain in Australia is the world's largest karst landscape, covering some 100,000 square miles (270,000 square kilometers).

Karst terrain is characterized by caves, sinkholes, disappearing streams, and underground rivers. Spectacular landforms may develop, as in the pinnacles of South China and the islands of Ha Long Bay in Vietnam. Groundwater in karst flows relatively freely through a network of solution-enhanced fractures, faults, and conduits. The water typically discharges at springs, including many of the largest springs on Earth. Karst aquifers are highly productive but extremely vulnerable to contamination.

The rocks that comprise the Edwards Aquifer formed during the days of the dinosaurs as reefs and lagoon deposits similar to the modern-day Bahamas. The hydrology of this particular aquifer is unusual, even for karst. Most of the water in the aquifer originates in Texas hill country, where streams flow over relatively impermeable rocks. Little water percolates into the ground until the streams cross a band where the faulted and fractured limestone of the Edwards Aquifer intersects the land surface. The streams then rapidly recharge the aquifer through fissures and sinkholes. As the water works its way down, it becomes confined between two relatively impermeable formations. Extensive faulting and dissolution along fractures results in a complex flow system through honeycombed limestone rocks. Eventually the water reemerges at springs where towns such as San Antonio, San Marcos, and Austin had their beginnings. But that's not the end of the story. The springs supply water to streams flowing southward to bays and estuaries along the Texas Gulf Coast, where coastal and marine wildlife depend on these freshwater flows.

The Edwards Aquifer is one of the world's most productive aquifers. In 1991, a "Texas-sized" well drilled into it was the world's greatest flowing artesian well, with a natural discharge equivalent to about a quarter of San Antonio's annual water use at the time. The owners, apparently thankful for their water blessing, named the well "Ave Maria No. 1." The plan was to raise 750,000 catfish on the same amount of water that San Antonio was using to support 250,000 human beings.[21] While the catfish were thriving, city residents were facing a new city ordinance that restricted lawn watering.

The fish farm closed after a year, but not because of water overuse. Rather, excessive amounts of fish urine were being discharged, without a permit, into the nearby Medina River. The farm briefly reopened in 1996, but eventually closed down for good. The San Antonio water utility paid more than $30 million to buy out the owner.[22]

There were two sides to the story. To San Antonio water officials, the catfish farm was an outrageous waste of a precious resource. In the eyes of many farmers, however, the well owner had been singled out for growing a food crop that seemed a much more responsible use of the water than keeping lawns green in the summer. In addition, the catfish farmer's use of water was almost entirely nonconsumptive, so the water was available for nonpotable reuse by others.[23]

In many ways, the Edwards Aquifer behaves more like a surface-water reservoir than an aquifer, quickly replenishing after large rainfall events. The Edwards is often compared to a bucket having different sized holes at several levels from top to bottom. When the aquifer is full, the water flows from all the holes (springs). Water discharge from each spring depends on its size and elevation. As the water level declines, the flows decrease. If the water level falls below the lower edge of its outlet, a spring dries up. The bucket-spring analogy is an oversimplified representation of this complex karst aquifer, but conveys the basic idea.[24]

Initially, the Edwards Aquifer was so prolific compared to demand that pumping had a negligible impact on the spring discharges. As a result of large groundwater withdrawals, however, today many of the springs rarely flow unless a flood fills the aquifer. Comal and San Marcos Springs near San Antonio (which are "lower in the bucket" than the dried-up springs) are now the major natural discharge points, and have become the center of huge controversies surrounding the aquifer.

Comal and San Marcos Springs are in exquisite settings at the edge of Texas hill country. Comal Springs, the largest in the American Southwest, consists of seven major springs and dozens of smaller ones at the base of a steep limestone bluff. San Marcos Springs consists of more than two hundred springs that issue from three large fissures and many smaller openings. Comal and San Marcos Springs have long been a popular draw.[25]

San Marcos Springs has a particularly colorful history. Human occupation goes back almost twelve thousand years, making it one of the oldest continuously inhabited places in North America. Early settlers described fountains that gushed water several feet into the air. Today, the springs lie at the bottom of Spring Lake, a reservoir created by a dam built to power a long-gone gristmill. Glass-bottomed boats provide tourists fish-eye views of the springs bubbling through sands into the clear lake.

In 1926, Arthur B. Rogers purchased land around Spring Lake with the goal of creating "one of the great playgrounds of Texas and the Southwest." The Aquarena Springs theme park opened in the 1950s, boasting what it called the world's only submarine theater. An audience seated in a large steel box was lowered into the crystal-clear spring water until its picture windows were almost entirely underwater. Ralph the Swimming Pig would then kick off the show with a "swine dive" into the water, enticed by an "Aquamaid" holding a baby bottle of milk. While breathing through hoses, Aquamaids then performed underwater acts, including ballet routines, drinking from soda bottles, and dining at a "bug-free" underwater picnic table with friendly fish invited to the "fishnic." Over the years, millions were entertained at Aquarena Springs.[26]

By the 1990s, the theme park's heydays had passed. The interstate highway bypassed the park, and SeaWorld San Antonio had opened in 1988. Aquarena was purchased by Texas State University at San Marcos in 1994. The university restored the site, with an eye to preserving its unique archaeological and biological resources. The switch from entertainment to education was highly controversial among the many people who had fond memories of family trips to Aquarena. Others would like to see Spring Lake drained and the springs restored to resemble their natural state.

Many people have a vested interest in the Edwards Aquifer. In addition to the tourist and recreational value of Comal and San Marcos Springs, the aquifer is a vital water resource. Municipal and industrial users in the growing metropolitan area depend on the Edwards for their water supply, competing for the water with families who have been farming the land for generations. The springs provide much of the flow of the downstream Guadalupe River, especially during droughts. Downstream surface-water users have a vested interest in maintaining flows, as do those concerned about the ecological requirements of freshwater flows to the Gulf Coast—winter home of the endangered whooping crane. The springs also help keep in check a so-called bad water line, which separates good quality groundwater from highly mineralized water. Too much pumping upgradient of the bad water line could result in saltwater intrusion into the freshwater aquifer. Finally, conservationists are concerned about the future of several endangered or threatened species that make the springs their home. This last use brought matters to a head.

In 1973, Congress passed the Endangered Species Act with the goal of protecting both species and their ecosystems. Virtually all species of plants and animals (except pest insects) are eligible. Under the act, endangered means that a species is in danger of extinction throughout all or a significant portion of its range, while threatened means that a species is likely to become endangered within the foreseeable future. The bottom line is that no federal agency may authorize, fund, or carry out any action likely to threaten or harm the existence of an endangered or threatened species. As interpreted by the U.S. Supreme Court, the intent of Congress in enacting the Endangered Species Act was to protect species from extinction at any cost. No one, public or private, can take an endangered or threatened species of fish or wildlife—with "take" broadly defined to include "harass, harm, pursue, hunt, shoot, wound, kill, trap, capture or collect."

The constant temperature and flow of high-quality waters of the Edwards Aquifer create one of the most diverse aquifer ecosystems in the world.[27] Among its wonders, blind catfish are occasionally pumped out of the aquifer from great depths. Seven endangered and one threatened species live in and around Comal and San Marcos Springs.

The Edwards Aquifer is extremely vulnerable to prolonged droughts. A "taking" of species listed under the Endangered Species Act occurs when the springs fall below critical levels required for aquatic habitat. During a severe drought in the 1950s, Comal Springs ceased flowing for more than four months. There is no record of the San Marcos Springs ever drying up, but its flow was greatly diminished during the drought. As withdrawals from the Edwards Aquifer increased, so did the possibility that the springs would again slow to a trickle or cease to flow altogether.

In 1991, the Sierra Club filed suit in the U.S. District Court in Midland, Texas. The lawsuit alleged that the Secretary of the Interior and the U.S. Fish and Wildlife Service had failed to protect endangered species by not restricting water withdrawals from the Edwards Aquifer. Concerned about maintaining river flows downstream of the springs, the Guadalupe-Blanco River Authority joined the lawsuit in support of the Sierra Club. Around this time, the Guadalupe-Blanco River Authority also unsuccessfully sought to designate the Edwards Aquifer as an underground river subject to surface-water laws.[28]

In February 1993, Judge Lucius D. Brunton III ruled in favor of the plaintiffs and required the U.S. Fish and Wildlife Service to determine the minimum spring discharge necessary to protect the listed species in both springs. Judge Brunton gave the state an opportunity to address the issue in the upcoming legislative session before "the 'blunt axes' of Federal intervention have to be dropped." If the Texas legislature did not adopt a management plan to limit aquifer withdrawals by the end of its current session, the plaintiffs could return to court. The Sierra Club indicated that, if it had to return to the District Court, it would play hardball by seeking federal regulation of the aquifer through the U.S. Fish and Wildlife Service.

On May 30, 1993, one day before Judge Brunton's deadline, the Texas legislature established the Edwards Aquifer Authority (EAA) to regulate groundwater withdrawals and manage the aquifer. A series of legal challenges ensued. In the end, a permit system based on historical use was established and is now administered by the Edwards Aquifer Authority.

The Edwards Aquifer Authority restricts total withdrawals, with further restrictions during dry periods. Although the endangered species forced the issue, many people have benefited from the proactive steps taken for responsible water management. As a result, the aquifer has continued to provide considerable water even during recent periods of extreme drought.

Saltwater intrusion is one of the most critical limiting factors in developing groundwater resources. The problem is most obvious along coastal areas where fresh groundwater comes into contact with the oceans, yet saline water also underlies freshwater in many interior aquifers. Freshwater is less dense and so tends to flow on top of the saline water.

When pumping occurs, the boundary (or transition zone) separating freshwater and saltwater moves toward the well. If the boundary moves far enough, the well will become saline, contaminating the water supply. Mixing in just 2 to 3 percent sea water (or equivalent saline groundwater) renders fresh groundwater unfit for most uses. Encroachment of saline water into a freshwater aquifer may be essentially irreversible, or take decades to be flushed out after the flow of freshwater has been reestablished. As a result, salinized wells are usually abandoned. Prevention of saltwater intrusion is by far the most effective strategy.

Saltwater intrusion is a serious problem in coastal aquifers throughout the world. In the United States, it was recognized as early as 1854 on Long Island, New York.[29] Saltwater intrusion in the Floridan Aquifer beneath Hilton Head Island, South Carolina, illustrates the challenges faced by many coastal communities today.

Underlying all of Florida and parts of Alabama, Georgia, and South Carolina, the Floridan Aquifer is one of the most productive aquifers in the world. The Upper Floridan aquifer has long served Hilton Head Island and is the principal source of water supply for coastal Georgia to the south, an area of rapid population growth. Historically, the aquifer discharged via springs at the north end of Hilton Head Island. These large offshore sources of freshwater (once well known to sea captains) disappeared as pumping reversed the groundwater flow direction from seaward to landward.[30]

The first well on Hilton Head went bad in 2000. Five more have since been lost, with saltwater intrusion continuing to advance beneath the island. It is expected that all Upper Floridan wells on the island eventually will be lost to saltwater intrusion.[31] Concentrated pumping near Savannah, Georgia (twenty miles from Hilton Head) has had the greatest effect on migration of the saltwater on Hilton Head Island, yet Savannah has not been so affected. This disparity has long been a sore point between the two states.

In recent years, Georgia and South Carolina have taken action to ameliorate the problem. Along with strong conservation measures, Hilton Head laid a pipeline to the mainland to import treated surface water, built a desalination plant to treat brackish water, and installed a well for recharging and later recovering excess water. For its part,

Georgia undertook scientific studies to develop a comprehensive plan to stabilize or halt saltwater intrusion in its twenty-four coastal counties. According to the Georgia plan, red zone counties (such as those near Hilton Head) have to reduce their pumping by five million gallons per day compared to 2004 levels. These measures will not solve the problem on Hilton Head, but will slow it down. In 2010, Georgia's Environmental Protection Division estimated that groundwater withdrawals would need to be cut by 90 percent in the Savannah–Hilton Head area to stop further saltwater penetration.[32]

The groundwater situation on the plush resort island of Hilton Head is serious, yet saltwater intrusion is even more problematic for many isolated Pacific islands that rely on shallow groundwater as their principal or sole source of water supply. There are about a thousand populated small islands in the Pacific Ocean. In many of these islands, groundwater resides in a very thin lens of freshwater surrounded by seawater. Highly sensitive to the effects of droughts and pollution, these aquifer systems are some of the most vulnerable on Earth. To make matters worse, their vulnerability to saltwater contamination arises not only from pumping, but also from storm surges or tsunamis. Climate change and a rise in sea level could greatly magnify these problems.[33]

In addition to saline water, freshwater resources exist around the world beneath the continental shelves. These large volumes of groundwater were deposited when the shelves were exposed to freshwater during sea-level lows (glacial maximums). Groundwater with a salinity of about one-quarter that of seawater has been found more than sixty miles (a hundred kilometers) offshore. Using that concentration as a benchmark, as much as a hundred times the global volume of groundwater extracted since 1900 may lie beneath the continental shelves. The problem is that accessing this huge resource is prohibitively expensive.[34]

Saltwater is viewed as a valuable resource in many interior water-scarce regions. Most of this interest centers on brackish groundwater—water that contains between 1,000 and 10,000 parts per million (ppm) total dissolved solids, a common measure of water salinity. By comparison, people drink groundwater that falls at the lower end of this range (1,000 to 3,000 ppm), but only if there are no other choices. The salinity of seawater is about 35,000 ppm.

About a third of the continental United States is underlain by brackish groundwater within three thousand feet (a thousand meters) of the land surface. On face value, this is an immense resource. In Texas alone, brackish groundwater is estimated at 2.7 billion acre-feet—enough to cover the state in fifteen feet (4.5 meters) of water.[35]

Much of the work characterizing saline groundwater in the United States was conducted during the 1960s and 1970s, when desalination was first being promoted as a future source of water supply. The interest was short-lived. Desalination was too expensive, and water availability was less of an issue than today. However, advances in desalination technologies and increasingly stressed water resources have revived interest in brackish groundwater. The world's largest inland desalination plant is in El Paso, Texas. An even larger facility is being planned for San Antonio.

Although the cost and energy requirements of desalination have been reduced, desalinized water is still the most expensive water on the planet. There are other drawbacks. Although saline groundwater might be considered a "new" source of water, the parts of aquifers containing saline water commonly are connected hydraulically to freshwater aquifers. In these situations, development of saline groundwater might affect the flow and quality of freshwater aquifers and surface-water bodies connected to the groundwater system. In contrast, developing brackish water aquifers that have little connection to shallower systems is more or less equivalent to mining a nonrenewable resource.

Water quality is also an issue. Brackish groundwater can have high concentrations of toxic elements such as arsenic and radium. It also has a greater tendency to precipitate calcite, sulfate, and other forms of scale that gum up the works of desalination plants. Then there's the challenge of how to dispose of the brine "waste product." Brines from desalination of seawater are pumped to the sea, while groundwater brine is typically disposed through evaporation ponds or reinjected deep underground.

While desalinated groundwater could increase the nation's water supply, much remains to be learned about the distribution and characteristics of brackish groundwater and the effects of extraction on fresh groundwater and surface-water resources. In addition, brine disposal remains a key environmental issue. As in the case of fresh groundwater, many factors beyond just the total amount must be considered in determining how much brackish groundwater is actually available for use.

6

Who Owns Groundwater?

The history of [groundwater law] is as thrilling as ignorance, inertia, and timidity could have made it.
—Mark N. Goodman

Who owns or has a right to use groundwater? How much can they use, and where can they use it? Can their water rights be sold? Establishing rules and laws to answer these questions has not come easy and remains controversial to this day.

Throughout most of human history, landowners dug wells without asking anyone's permission. In modern times, British common law established the precedent that land ownership carried with it absolute dominion over a "Blackstonian wedge" extending from the Earth's center up to the heavens.[1] As such, a landowner had the right to use the water under his or her land at any time and for any purpose. If a landowner's activity interfered with a neighbor's ability to withdraw groundwater from beneath his or her land, there was no legal redress.

This early hands-off approach reflected, in large part, the belief that groundwater was too mysterious to regulate. In 1861, the year the American Civil War began, a landmark Ohio court case ruled that groundwater was "so secret, occult and concealed, that an attempt to administer any set of legal rules would be involved in hopeless uncertainty, and would be, therefore, practically impossible."[2] The state clung to this view for more than one hundred years. In 1984, the Ohio courts finally ruled that groundwater use must be "reasonable" in comparison to the water needs and uses of one's neighbors.

Ohio's description of groundwater as being too "secret, occult and concealed" to regulate influenced many other states. Among these, Texas adopted what has become notoriously known as "the law of the biggest pump." Officially called the "rule of capture," landowners in Texas have the right to pump all the water they can from beneath their property, regardless of the impact on adjacent landowners. There are a few caveats. The landowner cannot waste the water or pump to willfully damage their neighbor. And they can't remove water by drilling a slanted well under their neighbor's land. (Local districts may set additional restrictions, as discussed later.)

In 1954, the Texas Court of Appeals upheld the rule of capture in a precedent-setting case pitting the Pecos County Water Control and Improvement District against Clayton Williams. Groundwater withdrawals by Williams and other irrigators were causing Comanche Springs, a treasured historical landmark and tourist attraction, to dry up. The court ruled that the water district had rights to the water only after it had emerged from the springs. Williams and surrounding irrigators could beneficially use any amount of groundwater underneath their land regardless of the impact on the springs.[3]

With its population expected to double over the next fifty years, Texas is acutely aware that its economic future rests as much on water as it does on oil and gas, a fact driven home by the recent multiyear drought that was among the worst on record. To its credit, Texas has one of the most comprehensive groundwater planning efforts of any state. Nonetheless, the state legislature and courts hold steadfast to the rule of capture.

To provide for some local control, the Texas legislature allows local Groundwater Conservation Districts (GCDs) to implement exceptions to the general rule of capture. GCDs can regulate the spacing of wells, limit pumping, and even deny a permit to withdraw groundwater. Since this authorization in 1949, nearly one hundred GCDs have been created—most over the past couple of decades.

Defined mostly along local political boundaries, the GCDs form a patchwork across the state. Two adjoining landowners may be drawing from the same aquifer but have to deal with entirely different regulations. In some cases, one landowner's pumping might be stringently regulated while his neighbor enjoys the generous rule of capture. In 2015, a classic such case came to a head in Hays County near Austin, Texas.

A water-supply company proposed to drill a well field in western Hays County from which water would be pumped from the Trinity Aquifer to Austin's fast-growing suburbs. It would be by far the largest commercial pumping project in the area. The Edwards Aquifer Authority regulates the overlying Edwards Aquifer, but has no jurisdiction over the Trinity Aquifer. The proposed well field was outside of any Groundwater Conservation District, falling within what lawyers call a regulatory "no-man's land." Consequently the company could pump as much as it wanted.[4]

Shallow residential wells, which provide water for most people in the fast-growing suburbs, had already gone dry during an ongoing drought. People feared that the proposed large-scale pumping would only make matters worse. Hundreds of residents responded with outrage in contentious community meetings. The office of the local state representative, Jason Isaac, was flooded with phone calls demanding action.

Representative Isaac took the matter to the Texas legislature. Six months later, after acrimonious debates, the legislature expanded the jurisdiction of an adjacent GCD to include the area where the proposed well fields would be drilled. But the conflict is far from settled. The GCD has little money to develop and enforce regulations, and any attempt to limit the company's pumping will likely end up in court.[5]

Groundwater Conservation Districts are reportedly the preferred method of groundwater management in Texas, but their authority remains tenuous in a state where the rule of capture still reigns supreme and groundwater is an inalienable property right. Two lawsuits against the Edwards Aquifer Authority (similar to a GCD) illustrate the challenges. The first case was filed by Burrell Day and Joel McDaniel, two farmers based south of San Antonio. It is known simply as the Day case.

In 1993, when the Texas legislature brought water rights in central Texas under the regulation of the Edwards Aquifer Authority, each landowner's water rights were limited to their historical groundwater use before the law went into effect. In 1996, Day and McDaniel requested a permit to pump seven hundred acre-feet (860,000 cubic meters) of groundwater per year to irrigate a peanut and oat farm on the ranch they had recently purchased. (Complicating matters, their plan was to use the free-flowing water from a well that had been drilled by the previous owner to fill a large man-made lake.) The Authority denied their request, awarding them only fourteen acre-feet of usage rights.[6]

Day and McDaniel took the Authority to court, claiming that they were being denied a property right to the groundwater beneath their land. The District Court ruled for the Authority, but its decision was reversed by the Court of Appeals in 2012. The case went all the way to the Texas Supreme Court, which summarized its decision in two sentences: "We decide in this case whether land ownership includes an interest in groundwater in place that cannot be taken for public use without adequate compensation guaranteed by . . . the Texas Constitution. We hold that it does." The case was the first explicit recognition by the Texas Supreme Court that landowners owned not only the water they pumped, but also the water remaining beneath their property. The decision resulted in considerable uncertainty about the ability of the Edwards Aquifer Authority, or a GCD, to restrict groundwater pumping without compensating the landowner.[7]

The second case was largely an extension of the first. Glenn and JoLynn Bragg, who had purchased their land years before the Edwards Aquifer Authority was established, sought permits for irrigating two pecan orchards. The couple had been irrigating one orchard with groundwater and the second by other means. They wanted to increase the groundwater applied to the first orchard and drill a well to irrigate the second. The Authority denied the permit for a new well and allowed only their historic water use (about half the amount requested) for the other. The couple alleged that limiting their water use was paramount to a "taking" of private property. The Braggs had planted the trees with the expectation of being able to use the water. There was an obvious economic hardship if they couldn't. The San Antonio Court of Appeals agreed, ruling that the couple should be compensated for the permit denial. The Edwards Aquifer Authority appealed this case to the Texas Supreme Court. In a surprise move, the Texas Supreme Court punted on the high-profile case, declining to review the Court of Appeals ruling.[8]

Texas landowners are permitted to sell, lease, or transfer groundwater rights to whomever they choose. Known as "water ranching," this practice was made famous by T. Boone Pickens, the legendary oilman known for his hostile takeover bids for large oil companies during the 1980s and 1990s. Unsuccessful in each bid, Pickens eventually lost control of his own company, Mesa Petroleum. Despite these setbacks, his entrepreneurial spirit remained fully intact.

In 1971, Pickens had bought property in Roberts County in the Texas Panhandle to diversify Mesa Petroleum into cattle operations. The diversification failed as the venture lost money, but Pickens held on to the property as a retreat and a place to hunt quail. The land overlies the High Plains Aquifer, but the gullies and hills of Roberts County are not conducive to irrigated agriculture. As a result, the county's large groundwater reserves remained mostly untapped.[9]

In the mid-1990s, the Canadian River Municipal Water Authority (CRMWA—often pronounced "crumwa") purchased groundwater rights near Pickens's ranch. Under the rule of capture, the utility could potentially suck water right out from under his ranch. Pickens, a high-rolling corporate raider who views water as a commodity no different than oil, was not about to let someone else sell *his* groundwater.

Pickens approached the utility, as well as the city of Amarillo, about purchasing his water. After both entities turned him down, he went on a buying spree, purchasing as many nearby groundwater rights as he could scoop up. A bidding war for the county's groundwater was soon in full swing with each party raising the stakes and trying to undermine the other. Within a few years, nearly 80 percent of the groundwater rights in Roberts County had been purchased by Pickens, CRMWA, and Amarillo.[10]

Pickens now owned more groundwater than any other individual in the United States, but he still had no buyer. He began to look further afield to the rapidly growing and thirsty cities of Dallas and San Antonio in East Texas. Pickens envisioned making a hefty profit by providing enough water for about 1.5 million Texans every day, while simultaneously keeping others from stealing his water. But there was a problem. The closest city, Dallas, would require a 250-mile (400-kilometer) pipeline and the rights to cross about 650 tracts of private property. The city considered the water too expensive and turned him down.[11]

If Dallas had agreed to buy Pickens's water, the power of eminent domain—the right of a government entity to force the sale of private property for the public good—would have come to the rescue. He needed eminent domain not only for a water pipeline but also for the transmission of electricity from his planned wind farm. A lot was on the line. Pickens had already invested about $100 million. Undaunted, he continued his quest.[12]

An opportunity for Pickens to garner the power he sought came in 2007, when the Texas legislature loosened the requirements for creating a water district. The five elected supervisors no longer needed to be registered voters living within the district—they only had to own land in the district. Pickens immediately saw an opening. He sold eight acres on the back side of his ranch to his lawyer in Houston, two of his executives in Dallas, and the couple who managed his ranch. A ballot initiative was placed for the November election to create an eight-acre (3.2-hectare) water district with the five new landowners serving as the directors and sole members. The couple, who managed his land and lived in the proposed water district, were the only people qualified to vote on the initiative. The vote was unanimous (2–0).[13]

With his new water district in place, Pickens not only had the power of eminent domain; he could also issue tax-free bonds to fund the project. But it all came to naught. Pickens was unable to find an

East Texas city as a buyer for his water and ended up selling his rights to his old nemesis, CRMWA.

The situation is somewhat different in the neighboring state of New Mexico, where groundwater is a public resource subject to prior appropriation (first in time; first in right) for beneficial use. The tighter restrictions on groundwater use can be seen from the air by a sudden drop in the density of green circles from center pivot irrigation as one crosses from Texas to New Mexico.

The doctrine of prior appropriation has its roots in the California gold rush. After the discovery of gold at Sutter's Mill in 1848, hundreds of thousands of people flocked to California to seek their fortunes. California had just been ceded to the United States by the Treaty of Guadalupe Hidalgo, which had ended the Mexican-American War. In the absence of new laws, the miners searched for gold and made their claims as trespassers. They weren't interested in land ownership. They wanted gold—to find it, take it, and move on to the next site.[14]

To settle land disputes, the miners set up vigilance committees and adopted the most elementary form of ownership—the first to grab it, owns it. In more civilized terms, this became "first in time, first in right." It was only natural to apply this same principle to the water that was essential to work a claim. Thus prior appropriation became established for regulating surface water in all western states. Beginning with New Mexico in 1931, most western states extended the prior appropriation doctrine to groundwater. A legal system that "arose from the relatively lawless mining camps of the Wild West would come to be viewed as though it had been handed down directly from God," observed University of New Mexico professor Reed Benson.[15]

For surface water, the priority system adjusts everyone's appropriations in response to fluctuating stream flows so that senior water rights are protected. The protection of senior water rights for groundwater is much more difficult. In this case, priority can be addressed, in part, by limiting the number of well permits in order to protect established users from excessive declines in groundwater levels. As we will see in the next chapter, the interconnectedness of groundwater and surface water further complicates matters. There is also the question of how to deal with the "little guy"—the homeowner solely dependent on a private well for water supply. As water resources have become stretched, New Mexico and other western states have become a battleground over this issue of regulating domestic wells.

In New Mexico, the permit application for a new well must show that the well will not impair existing water rights. If the well is for household and other domestic uses, however, the permit is automatically approved. Given the relatively small use, the state considers it too time-consuming and burdensome to evaluate each application. These are called "exempt wells," and most western states have similar provisions.

Although it is true that domestic wells generally use little groundwater compared to irrigation or municipal wells, collectively the amount of water pumped adds up. New Mexico has over 200,000 legally permitted domestic wells, with many more drilled illegally.[16] In 2006, Horace Bounds, a farmer in the Mimbres Basin in southwestern New Mexico, challenged the exception for domestic wells in court. While the Mimbres Basin had been closed to further groundwater development for decades, new homes with domestic wells continued to be built. Bounds alleged that these well permits violated his water rights dating back to 1869.

The New Mexico Supreme Court ruled that the act of issuing a permit did not violate Bounds's rights, but went on to stipulate that the state has a legal obligation to defend the rights of senior water users, including curtailing domestic water use if necessary. In other words, issuing a permit to drill a well doesn't give the owner unlimited permission to use it. "It puts anyone getting a new domestic well on notice," observed State Senator Peter Wirth.[17] The issue of exempt wells remains controversial throughout the west.

In addition to the exemptions for shallow domestic wells, for many years New Mexico provided a carte blanche exemption to its permitting system for wells in brackish aquifers deeper than 2,500 feet (760 meters). Only a "notice of intent" (no permit) was required for these wells. Few such notices were filed until a veritable gold rush occurred around 2006. With growing water scarcity, cities, developers, and industries suddenly all wanted to stake their claim to this untapped resource. As much as 75 percent of New Mexico's groundwater is brackish, and the state may contain one billion acre-feet of it.[18]

Within a few years, notices were filed for more than six hundred wells with appropriations of over 1.7 million acre-feet (two billion cubic meters) of brackish groundwater—about the same as New Mexico's current total groundwater withdrawals.[19] Potential effects of this development on connected fresh groundwater and surface-water resources raised a red flag. In response, the state revised its statutes and empowered the state engineer to regulate the brackish groundwater resource through permits. The story is far from over. Many uses are exempt from this regulation. It's also not clear how much brackish groundwater can be developed without impairing existing water rights.

California presents yet another variation on who owns groundwater. The state's water resources largely depend on precipitation that falls during some five to fifteen days each year, making groundwater an essential resource. Most years, groundwater accounts for about 40 percent of the state's water supply; during droughts that percentage increases to as much as 60 percent. Much of this water goes to agriculture. Along with the High Plains, California was the first region on the planet to use large amounts of groundwater for irrigation. In 1908, the term "overdraft" was coined by W. C. Mendenhall in reference to groundwater depletion in southern California.[20]

Until as recently as 2014, California was the only western state without a comprehensive statewide groundwater management program. With a few exceptions, groundwater management was viewed as a voluntary local activity. Conflicts in some groundwater basins ended up in court, where withdrawal rights were assigned to each user. Known as "adjudicated groundwater basins," this costly and lengthy process is considered a last resort. By 2015, there were twenty-three adjudicated basins in California, nearly all located in southern California.

The state's groundwater history goes back to 1903, when the California Supreme Court rejected the English common law doctrine that landowners have absolute rights to the groundwater beneath their land. Instead, multiple owners above an aquifer each possessed a "correlative right" with proportional sharing of groundwater tied to their land ownership. Any "surplus" groundwater not needed by overlying landowners could be acquired by others through the doctrine of prior appropriation. In addition, groundwater rights could be acquired by individuals who went unchallenged in taking more than their fair share. This confusing mix led to a legal free-for-all, contributing to unchecked pumping and overdrafted aquifers.[21]

In 1949, the State Supreme Court sought to inject some order into this chaotic situation by creating the doctrine of "mutual prescription" for an overdrafted aquifer. Water rights to the aquifer were calculated by determining each user's highest five years of pumping, then proportionately reducing their allotment until the total did not exceed the presumed safe yield of the aquifer. This policy shift sparked a "rush to the pump house," as people sought to set their average uses as high as possible.

The severe drought of 1976–1977 drew attention to the need for better water management. Governor Jerry Brown appointed a commission to review California water-rights law, in part to address groundwater problems. The commission recommended a number of actions, but the political system lacked the will to adopt them. No action was taken.[22]

In 1992, a pared-down version of the proposed legislation by Governor Brown's commission was passed as the Groundwater Management Act (AB 3030). The act provided the authority and guidelines for local agencies to manage groundwater. Over the next decade, numerous water agencies developed groundwater management plans, but there were no reporting requirements. In addition, local agencies were not required to keep their plans up-to-date, or even to implement them. Many early plans focused more on preventing the export of groundwater from their area than on management within the area.[23]

For the next two decades, local agencies incrementally strengthened their groundwater management with a patchwork of restrictions.

Progress was very slow. Meanwhile, opposition from powerful agricultural interests (and fragmented state agencies with too much on their plate) all but nullified groundwater management at the state level.

A change in mindset began with the onset of the drought of record. The year 2013 was the driest year since the state had started measuring rainfall in 1849, and record heat intensified the impact. The width of old tree rings suggested that California had not been this dry for five hundred years.[24] Seventeen rural communities in the state were in danger of running out of water within sixty to 120 days.

In January 2014, with surface-water reservoirs at historic lows, the State Water Project announced that it would cut off water to local agencies serving twenty-five million residents and about 750,000 acres (300,000 hectares) of farmland. The Metropolitan Water District, serving much of southern California, gets about 30 percent of its water from the State Water Project. The following month, the Bureau of Reclamation announced that farmers should also expect no irrigation water from the federally run Central Valley Project.[25]

The stakes could hardly have been higher for the country's most populous state and its $45 billion agricultural industry. California produces nearly half of all the fruits, nuts, and vegetables grown in the United States. It's also the leading dairy state. California farms depend heavily on irrigation to sustain this production. A large part of the water deficit from the cut-off surface-water supplies would come from increased groundwater pumping of the state's already overdrafted aquifers.

Heedless of all consequences, a drilling frenzy ensued. Farmers with older, shallower wells watched them go dry as neighbors drilled deeper and sucked the water table down. Some farmers went into substantial debt as they engaged in a drilling "arms race" that they could not afford to lose. Pumping costs rose as water levels dropped. Despite being the third year of record drought, farmers managed to idle only about 5 percent of the state's irrigated land—but at substantial cost to its groundwater resources.[26]

In a curious twist of fate, Jerry Brown (who had been governor thirty-six years earlier during the failed attempts to regulate groundwater) was back in office. In December 2013, Governor Brown released the California Water Action Plan, which outlined proposed actions for the next five years. Improving groundwater management was, once again, a key priority. But this time, the governor managed to get people's attention with this statement: "When a basin is at risk of permanent damage, and local and regional entities have not made sufficient progress to correct the problem, the state should protect the basin and its users until an adequate local program is in place." The state was no longer going to sit idly by and watch the groundwater debacle continue to unfold. The next month, Governor Brown's proposed budget included funds for ten additional State Water Resources Control Board staff members "to act as a backstop when local or regional agencies are unable or unwilling to sustainably manage groundwater basins." The governor was serious about California's need for groundwater management, and everyone knew it.[27]

The momentum continued to build. State Senator Fran Pavley of Agoura Hills and Assemblyman Roger Dickinson of Sacramento each introduced bills for statewide groundwater management. The Association of California Water Agencies (ACWA) put forward a similar set of recommendations that largely echoed those of Governor Brown's previous commission in 1978. ACWA is the largest statewide coalition of public water agencies in the country, with 430 water agencies accounting for 90 percent of water deliveries to cities, farms, and businesses in California. Limits and pump taxes were part of the ACWA plan. The California Water Foundation followed with similar proposals in support of statewide groundwater management. The Groundwater Resources Association of California, a professional organization of hydrogeologists, and the University of California at Davis became actively engaged in educating the public and legislators on the state's groundwater. Editorials and news articles proliferated statewide. The governor's office held almost daily meetings on the drought and the proposed legislation. It was a remarkable turn of events.[28]

The support was driven home not only by the current drought, but also by projections that climate change will cause major reductions in the Sierra snowpack that California depends on for much of its water supply. For the first time, a large cross-section of the population had become aware that the state's water future was tied to its groundwater resources. The California Farm Bureau Federation (and a few others) continued to oppose legislation, but were staunchly outnumbered. Everyone was talking about groundwater and, for once, they were all saying the same thing.

On September 16, 2014, Governor Brown signed into law the Sustainable Groundwater Management Act (SGMA, pronounced "sigma"). The law focuses on groundwater basins designated as medium or high priority by the California Department of Water Resources. There are more than a hundred such basins, and they account for most of the groundwater used in California that is not already adjudicated or managed by special districts. The majority of the basins are in the San Joaquin Valley—the state's agricultural heartland.

Under SGMA, the first step is for local agencies to designate a groundwater sustainability agency (or agencies). Each agency has five to seven years to adopt a plan that puts the basin, or subbasin, on track toward "sustainable management" by roughly 2040. Each plan must include monitoring and management over a fifty-year planning horizon, and measurable objectives must be achieved every five years. The groundwater sustainability agencies are authorized to limit groundwater pumping, monitor water withdrawals, and assess regulatory fees to fund groundwater management and replenishment activities.

Only through adoption and implementation by stakeholders will effective management be achieved. In enacting SGMA, the California legislature sought to manage groundwater basins through the actions of local governmental agencies "to the greatest extent feasible." If the locals can't agree on a designated groundwater sustainability agency, or if a suitable sustainability plan is not prepared within the designated timeframes, the State Water Resources Control Board can develop and enforce its own plan.

Many issues have yet to be resolved, beginning with the fact that "sustainability" is a nebulous concept. The act also left the controversial topics of water permitting, requirements for metering wells, and reporting groundwater use to the local agencies. Finally, there's the big question on which everything depends: will communities rise to the occasion and police themselves through both wet years and dry years? Formulating laws and regulations through the Sustainable Groundwater Management Act is the first step, and an important one. The proof, however, will be in the pudding—not in the recipe.

The challenges of groundwater management play out differently around the world. The European Union has established an overarching framework for water resources known as the Water Framework Directive. The directive places a high priority on protecting groundwater. This emphasis is not surprising given that groundwater provides approximately 70 percent of Europe's drinking-water supply. Austria, Croatia, Denmark, Hungary, Italy, and Montenegro obtain all or most of their drinking water from groundwater.[29] The importance of groundwater to the environment is also recognized by the directive. Water is not just another commercial item—it is a heritage that must be protected and defended. Yet such changes in attitude and practices are not easy. Spain, where groundwater has been used intensively for irrigation, is a case in point.

Groundwater was long held as a private property right in Spain, dating back to the country's 1879 Water Act. Major changes came with 1985 reforms to the Water Act, which declared groundwater to be public property. To avoid having to pay existing well owners for their groundwater rights according to Spain's Constitution, the government offered various benefits to encourage well owners to cede their rights. All well owners had to register their wells and choose between private and public ownership within three years. Water authorities could impose fines to enforce this rule. Despite this, perhaps 80 to 90 percent of wells remain undeclared to this day, as just one example of what has been called "hydrological insubordination."[30]

The 1985 Water Act emphasized self-regulation, with users of overdrafted aquifers required to organize themselves into groundwater user associations. This idea was a natural extension of Spain's long history of community decisionmaking for managing surface-water irrigation systems. For example, the Tribunal de las Aquas de Valencia

(Water Jury of Valencia) has been meeting for centuries every Thursday at noon at the entrance to the Cathedral of Valencia. Dating back over a thousand years to the time of the Moorish conquest, it is the oldest democratic institution in Europe. All members of the tribunal are elected by farmers and have equal standing. The judges sit in a circle in full public view where they resolve any disputes involving a nearby surface irrigation system in a swift and down-to-earth manner. The court is purely oral. Nothing is written and no records are kept. The system works.[31]

Despite this history, implementation of effective groundwater user associations has been difficult to achieve. Farmers dependent on groundwater for irrigation have much greater control over their water use, and a greater sense of entitlement to the resource, than those dependent on surface-water systems that are operated by a central authority. Consider La Mancha, a semi-arid plateau region in central Spain immortalized as the home of Don Quixote, the fictitious adventurer who set out to revive chivalry accompanied by his faithful squire, Sancho Panza. Don Quixote's relentless attacks on windmills are a clue to the region's agricultural heritage of barley, wheat, and other grains.

During the past forty years or so, the availability of groundwater and its reliability during droughts has brought significant benefits to La Mancha through the irrigated agriculture of grains and higher cash crops. At the same time, this intensive use of groundwater has dried up wetlands and river reaches. Among these, Las Tablas de Daimiel (The Wetlands of Daimiel) is a national park and the core of the La Mancha Húmeda wetland area, which has been declared a Biosphere Reserve by UNESCO. The wetlands are important to the survival of many migratory birds.[32]

For many years, the wetlands were drained by farmers to create more arable land. Conservation groups successfully managed to largely eliminate this practice in the 1970s, but deterioration of the wetlands from intensive use of groundwater went unchecked. In the 1980s, the flow of water to Las Tablas de Daimiel dried up and the water level fell below the bottom of the wetlands. The wetlands dried to the point that the peat in the soil would spontaneously catch fire, sometimes burning underground for long periods of time as smoldering fires propagated through the subsoil peat layers.[33]

In 1987, with the new Water Act in place, the Guadiana Water Authority (the basin authority responsible for groundwater management) declared the Western La Mancha Aquifer officially overdrafted. Construction of new wells was forbidden and pumping quotas were established. Instead of complying with the new rules, however, thousands of farmers simply ignored them. Illegal pumping became rampant. With few resources for enforcement, the drilling ban led to a sense of impunity among users, many of whom took advantage of the situation to install new wells to convert dry agricultural lands to irrigation. A severe drought in the 1990s further catalyzed the rush to drill new wells.[34]

In 2005, the basin authority brought the owners of five thousand illegal wells to court. Yielding to pressure from farmers and the local government, the federal Ministry of Environment fired the president of the basin authority and its water commissioner. The wetlands continued to deteriorate and major fires broke out in 2009.[35]

The Western La Mancha Aquifer illustrates the difficulty of developing groundwater regulations after water users are firmly entrenched. Without their cooperation, it is nearly impossible to achieve good groundwater governance. The basin authority for the Western La Mancha Aquifer tried to impose regulations from the top down without the agreement of farmers. In response, the groundwater user association largely opposed the basin authority and became a lobbying force for the farmers, even hiring lawyers to defend the farmers in court. To curb the overdraft, the government plans to spend 5.5 billion euros over a twenty-year period in order to purchase water rights and employ other conservation efforts. Unfortunately these funds have dried up in recent years because of Spain's economic woes.[36]

Spain is not the only country where the nation owns the water but has limited control over it. Numerous illegal wells exist throughout North Africa and the Middle East, despite the fact that all water resources are owned as a public resource or held in trust by each country.

In Australia, the driest inhabited continent, water is a public resource with allocations historically determined by the states (and territories), each with independent water laws and policies. The Australian states give "entitlements" to withdraw groundwater or surface water by issuing a license for a fixed period. (Most licenses are automatically renewed.) Depending on overall availability, water is allocated on a seasonal basis against these entitlements. Some states differentiate between high-reliability and low-reliability water entitlements, with the former receiving priority. In contrast to the prior appropriation system in the United States, shortages are shared among users.[37]

From 1997 through 2009, Australia's water resources were severely tested during a period of unprecedented drought known as the Millennium Drought—or simply "The Big Dry." Australia's economy and environment were hit hard. Reservoirs for cities were nearly empty throughout the country. Many farmers faced financial ruin. Dairies in South Australia sold off their entire herds of cows. Thousands of drought-resistant red gum trees died, many of them hundreds of years old. By March 2007 (late summer in Australia), the drought had become so severe that camels in Australia's Northern Territory were dying of thirst.[38]

In April 2007, Prime Minister John Howard, who had served from the outset of the Big Dry, warned that farmers would not be allowed to irrigate their crops at all the following year unless unexpectedly heavy rain fell in the next few months. He asked the country to pray for rain. The rain came, but not enough to stave off the drought or rescue Howard's political career. Howard lost the election that year, in large part because of his lack of action on global warming—a major election issue in the midst of a seemingly never-ending drought.[39]

Even before the onset of the Big Dry, Australia's water resources were overallocated. Water reforms began in the 1990s, and were formalized

The skeletons of Australia's iconic Red Gum trees haunt the shrinking shores of Lake Pamamaroo, near Menindee, New South Wales. *Source:* J. Carl Ganter/ Circle of Blue (circleofblue.org). Used with permission.

as a National Water Initiative in 2004. Referred to as Australia's "enduring blueprint for water reform," this ambitious agreement between the federal and state governments has among its objectives "the return of all currently over-allocated or overused systems to environmentally-sustainable levels of abstraction."[40] Prior to these water reforms, many wells were neither licensed nor measured. The National Water Initiative began to change this. The connections between groundwater and surface water also were acknowledged by the initiative, with commitments to a "whole of water cycle" approach.

A particularly noteworthy aspect of Australia's water reforms has been the emphasis on water markets, whereby farmers can trade their water allotments for a season or sell their entitlements altogether. Water markets create flexibility for users. Government has a role in the development of market rules and principles, but is not involved in choosing among competing potential water users. The economic benefits from water trading include optimizing the overall use of water in overallocated markets by promoting shifts from lower to higher value uses. There are also potential pitfalls to water markets. Rules are needed to ensure that third-party effects are controlled, and that the market for water is not cornered by speculators.[41]

Water markets made a huge difference in the economic survival of farmers during the Big Dry. Orchards and vineyards purchased water from dairy owners and rice farmers to keep their trees and vines alive. Dairy owners used the payments to purchase, rather than grow, their own fodder. Payments to rice farmers allowed them to fallow their fields without risking financial ruin. Despite a two-thirds drop in agricultural water use during the Big Dry, farm revenues dropped by only about 20 percent. When the drought was over, Australia's agricultural production quickly recovered.[42]

The Murray-Darling Basin in southeastern Australia was the center for water reform. Spanning four states (Queensland, New South Wales, Victoria, and South Australia) and the Australian Capital Territory, the Murray-Darling is Australia's "food basket" and accounts for about 85 percent of irrigated agriculture. The basin also provides the water supply for cities and towns, including Adelaide.

Even before the drought, the Murray-Darling Basin was not healthy. Expansion of irrigation for over a century had resulted in a vastly modified and overallocated river-aquifer system. Wetlands had disappeared, and the salinity of the area's rivers had increased due to the flushing of salt from the basin's naturally saline soils. The salinity affected not only irrigated crops but also tap water in Adelaide and other urban areas in South Australia. The sluggish rivers were also increasingly contaminated with fertilizers and pesticides. Fish, including the iconic Murray cod, were in decline.

In 1991, an algal bloom choked a 625-mile (thousand-kilometer) stretch of the Darling River, killing just about everything that lived in the river, tainting public water supplies, and causing New South Wales to declare a state of emergency. The toxic bloom mobilized the environmental community to take action and greatly heightened awareness of the poor condition of the entire river system.

Droughts have long plagued the Murray-Darling Basin, but the Big Dry was beyond anyone's imagination. By 2002, the Murray River ran so low that the mouth of the river silted up and had to be dredged to reach the sea.

Over the years, the states have made various attempts to jointly manage the Murray-Darling. In 1915, New South Wales, Victoria, and South Australia signed an agreement on how to share the Murray River. In 1997, a cap was placed on surface-water diversions from the Murray-Darling River system at around 1993–1994 levels. This level, however, was not based on any scientific assessment of the environmental water needs of the basin. In addition, groundwater use continued to grow, causing an unaccounted-for depletion of surface water. The states have a history of squabbling over the river, with upstream states pitted against downstream states and agricultural interests versus cities. As the drought wore on, disagreements intensified and more drastic action was needed.

In 2007, on the eve of Australia Day (January 26—the day celebrating the arrival of the first fleet of eleven convict ships from Great Britain in 1788), Prime Minister John Howard declared the management of the Murray-Darling broken. He demanded that the states hand over their powers to the Commonwealth, so that it could address, once and for all, environmental degradation of the Murray-Darling Basin and place a sustainable cap on surface water and groundwater use. This increased federal government involvement would be accompanied by significant funding. To get buy-in from the states, the Commonwealth committed 12.9 billion Australian dollars over ten years to modernize the irrigation infrastructure and buy back water from farmers.[43]

Later that year, the Commonwealth Water Act established the federal Murray-Darling Basin Authority and charged it with developing a plan for managing the basin's water resources. This complex and controversial undertaking was developed essentially behind closed doors, without local input. When a guide to the proposed plan was released in 2010, farmers were enraged at the large reductions in water use called for by the proposal. In at least one rural town, angry mobs set off large bonfires fueled by copies of the draft document. Australians watched mesmerized as these scenes were televised around the world. After the backlash, the Basin Authority brought in a new leader to engage and mend fences with the public. The final Basin Plan was approved by the Australian Parliament in 2012 without much fanfare.

The Murray-Darling Basin Plan introduced aquifer-by-aquifer limits on groundwater extraction—although these do not take effect until 2019. During the plan's development, limits on groundwater extraction bounced around and ended up much more generous than those proposed in the ill-received 2010 plan. New South Wales, in particular, received much higher allocations, possibly to ensure access to future water supplies by coal-seam gas producers and other mining operations.[44]

Australia's groundwater management remains a work in progress. The country may be backsliding to some degree as the Big Dry recedes into memory. Nonetheless, groundwater management in Australia has several noteworthy features—licensing and metering (albeit, still limited) of groundwater pumping, development of water markets and trading, an ability to share shortages, and joint water management by the states and Commonwealth.

Returning to the original question of who owns groundwater, it clearly depends on where you live. It's also usually a question of who has what rights, rather than absolute ownership. Yet even in cases where groundwater is a public resource, individuals typically perceive that groundwater is theirs to use as they please.

7

Streamflow Depletion

Nothing motivates like a crisis (and a Supreme Court Decree)
—Greg Lewis

The Skagit River originates in Canada and flows through the North Cascade Range of Washington State into Puget Sound. Although most people have never heard of it, the Skagit is among the largest rivers discharging to the West Coast of the United States (outside Alaska). The Skagit is the sole river system in Washington that supports all five species of salmon, not to mention grizzly bears, bald eagles, and trumpeter swans. But the salmon are the center of controversy.

For the Swinomish tribe, the salmon are a way of life. In 1855, the chiefs of the Swinomish and other tribes gathered at Múckl-te-óh (present-day Mukilteo, Washington) to sign the Treaty of Point Elliott. The treaty gave vast tracts of land to white settlers, but reserved the tribes' right to fish on their accustomed fishing grounds. After more than a century of off-and-on controversy, including fish-in protests by the tribes in the 1960s and 1970s, the U.S. District Court ruled that the tribes were entitled to half of all salmon that ran through their native waters. The decision was later upheld by the U.S. Supreme Court.[1]

Salmon need sufficient water at a low enough temperature to survive. In 2001, the Washington Department of Ecology issued an instream flow rule for the Skagit River based on a recently completed study of salmon water needs. The fish now had water rights requiring minimum flows to be maintained during droughts and low-flow periods.

The salmon had priority over any future water development. When streamflow is lower than that needed by the fish, people with more junior water rights could be cut off. In addition, if a basin's water had been fully appropriated, new water diversions would be prohibited. This prohibition would include wells, recognizing that pumping reduces surface-water availability.

Skagit County challenged the rule in court. Although the county had helped fund the study behind the in-stream flow rule, officials said they made a mistake in accepting an agreement that left out water for new rural well users. In 2006, the Washington Department of Ecology agreed to amend the rule to allow for limited "reservations" of surface water and groundwater for future uses in twenty-five basins within the Skagit watershed. A small number of new homes that rely on well water could be built, even in fully appropriated basins. The reservations represented less than 0.5 percent of main stem flows during low-flow conditions and were limited to 2 percent of low flows in individual tributaries. The Department of Ecology argued that these relatively modest water reservations would not significantly harm fish and wildlife.[2]

The Swinomish tribe disagreed. In 2013, the state supreme court struck down the Department of Ecology water reservations. The rights to groundwater were invalidated for an estimated 475 homes and eight businesses that had been constructed since the water supply had been allocated. The Department of Ecology and the Swinomish tribe agreed not to curtail water use for these well owners while mitigation solutions are sought. Meanwhile, those who had purchased land with the intent to develop it were caught in limbo in obtaining a building permit. The Department of Ecology is evaluating various ways to offset well-water use, such as the purchase and retirement of other water rights in the basin, as well as water storage and recharge projects.[3]

This battle over water is perhaps surprising in the wet climate of western Washington, yet it is just one of a growing number of cases where the connection between surface water and groundwater has become the center of controversy. To scientists, the interrelationship of groundwater and surface water is well established. What's true in nature, however, has long been ignored in courts and legislatures where separate laws and regulations govern groundwater and surface-water use. As surface-water bodies have become fully allocated, it has become incrementally more difficult to ignore the reality that groundwater and surface water are a single resource.

Upton Sinclair, who exposed the inhumane conditions in the meat-packing industry of the early twentieth century, famously observed, "It is difficult to get a man to understand something, when his salary depends on his not understanding it." By the same token, having allocated water without recognizing the connection of groundwater and surface water makes it difficult to admit that water has been overallocated.

Groundwater is far more than just a static pool of underground water. The term "aquifer" comes from the Roman "aqua ferre," meaning "to hand or carry water." Aquifers carry water underground from areas of recharge to areas of discharge along flow paths that range from a few feet to hundreds of miles. Velocities are typically inches to feet per day. Most groundwater eventually ends up as baseflow in streams, which explains why streams continue to flow during a prolonged absence of precipitation.

Groundwater is part of the hydrologic cycle—the most fundamental concept in hydrology. The hydrologic cycle is continuous, with neither a beginning nor end. It is often presented as a simple diagram showing evaporation of water from the sea, precipitation falling on the land, and water flowing to the sea by surface and subsurface (groundwater) routes. In reality, this larger cycle is short-circuited many times as water returns to the atmosphere before reaching the sea.

The hydrologic cycle was long misunderstood. Early Greek philosophers hypothesized that springs and streams were formed by seawater that moved through subterranean channels below the mountains and then, through some mysterious mechanism, was raised to the surface. This early view is essentially the reverse of what actually happens. To the Greeks, it was inconceivable that rainfall was sufficient to support

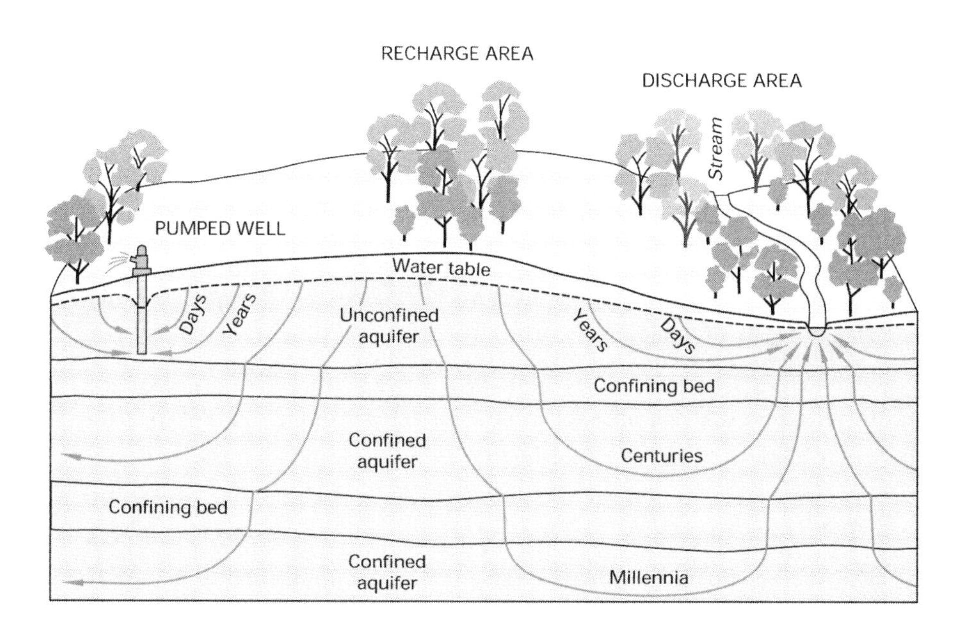

Groundwater flow paths vary greatly in length, depth, and travel time from recharge to discharge. *Source:* U.S. Geological Survey Circular 1139.

springs and rivers. Apparently, the question of how the ocean water lost its salt on its path from the sea did not trouble them.[4]

Roman architect Marcus Vitruvius, best known for his contributions to architecture and the acoustics of buildings, was one of the first to correctly grasp the hydrologic cycle. During the first century BCE, Vitruvius taught that water from melting snow seeped into the ground in mountainous areas and reappeared at lower elevations as springs. His writings were forgotten with the collapse of the Roman Empire.

For nearly two thousand years the hydrologic cycle remained misunderstood. English mathematician and philosopher Alfred North Whitehead succinctly expressed the overall lack of scientific progress: "In the year 1500 A.D. Europe knew less than Archimedes who died in the year 212 B.C." In medieval Europe, all knowledge had to be reconciled with Scripture. Accordingly, the biblical passage from Ecclesiastes "All the rivers run into the sea; yet the sea is not full; unto the place from whence the rivers come, thither they return again" was interpreted literally to mean that there must be direct circulation from the sea to land.[5]

It took three Frenchman and an Englishman to finally dispel these erroneous views. The first was Bernard Palissy (ca. 1510–1589), a surveyor and mad potter who once burned his garden fence, household furniture, and floorboards to heat his kiln in search of a special white glaze that would bring him riches. Palissy knew no Latin or Greek, so he was largely ignorant of the scientific theories of the day. As a result, he relied on his own observations of how water moves. "I have had no other book than the sky and the earth," he explained. Palissy believed that water was cycled out of the oceans by evaporation, followed by condensation into rain and snow. He published his theory, but it was ignored by the day's influential scientists who continued to expound theories similar to those of the ancient Greeks.[6]

Pierre Perrault went to the trouble of actually measuring rainfall in the upper Seine River Basin over a period of three years (1668–1670). He discovered that precipitation on the basin was about six times his estimate of river discharge, thereby disproving the Greeks' assumption of inadequate rainfall. Perrault was among the first to recognize that river flow, in the absence of precipitation, is maintained by groundwater discharge. Despite his accomplishments, Pierre was overshadowed by his younger brothers—most notably Charles, whose Mother Goose fairy tales brought us such classics as Little Red Riding Hood and Cinderella.[7]

Physicist Edme Mariotte (ca. 1620–1684) confirmed Perrault's work through measurements of the Seine River at Paris. Finally, the English astronomer Edmund Halley (1656–1742) essentially clinched the concept when he demonstrated that evaporation from the Mediterranean Sea sufficiently accounted for all the river water flowing into the Mediterranean. The rudiments of the hydrologic cycle were finally well established.[8]

Though often misunderstood, groundwater recharge and discharge are essential parts of the hydrologic cycle. Recharge is a capricious process, both highly variable across landscapes as well as episodic. It is also difficult to quantify. Most precipitation that infiltrates into soils doesn't make it all the way to the water table as recharge. Instead, it is returned to the atmosphere via evaporation and plant transpiration, collectively known as evapotranspiration.

Many factors influence the percentage of precipitation that recharges groundwater. Natural factors include previous precipitation, storm duration and intensity, soil and geologic properties, vegetation, and topography. Following a long dry period, even a large storm can result in little or no recharge as the water merely fills up a moisture deficit in the subsurface above the water table. As the aridity of a region increases, very little or no recharge may occur directly from precipitation and localized recharge from surface-water bodies becomes more important.

Humans affect recharge through irrigation, clearing of native vegetation, urbanization, and many other activities. Surprisingly, urbanization often increases recharge, as runoff from built-up and paved areas is channeled to a retention basin or other recharge area. Leaky water mains and sewers also contribute to recharge.

Groundwater discharge, the outflow of water from the groundwater system, occurs naturally at springs and as flow to streams, lakes, wetlands, and estuaries. Groundwater discharge also occurs as evapotranspiration by plants known as phreatophytes, which get their water from below the water table. Finally, humans contribute to groundwater discharge through wells.

The balance between groundwater recharge and discharge controls the amount of groundwater being stored in the same way that deposits and withdrawals control the amount of money in a bank account. For example, when recharge exceeds discharge, groundwater storage increases. Under natural conditions, a groundwater system is in long-term equilibrium—the amount of water recharging the system is approximately equal to the amount of water leaving (discharging) from the system. Pumping groundwater changes this balance.

At this point, we need to take a slight detour to explain two key concepts: "head" and "capture." Head refers to the energy available to move groundwater at a specific location. A water-level measurement in a well under static (nonpumping) conditions provides a measure of the head in the aquifer at that location. Groundwater flows from locations of higher head to those of lower head. The rate of change in head in a given direction is called the "hydraulic gradient." As hydraulic gradients in a groundwater system change in response to pumping, the magnitude and directions of groundwater flow also change.

The most obvious source of water to a pumping well comes from groundwater storage, causing water levels (heads) to decline. A second source, known as capture, comes from the environment in the form of less water for plants and surface water. Capture occurs as water-level (head) declines spread out from a pumping well. If pumping lowers water levels (heads) near springs, for example, the flow of those springs will decrease. In this case, the capture is equivalent to the amount that spring flow has been reduced by pumping. The most common form of capture is streamflow depletion, which results when pumping either causes reduced discharge to the stream (similar to captured spring flow) or induces a flow of water from the stream to groundwater. Similar effects may occur for a lake or wetland. Another form of capture comes when water-table declines cause plants to die or otherwise reduce the transpiration of groundwater. In this case, capture equals the reduction in transpiration caused by pumping.

As a well is pumped, the amount of groundwater coming from storage versus capture varies over time. The simplest example is a well near a stream that is pumping at a steady rate. Initially, pumping just removes water from storage in the aquifer, but as pumping continues, the water-level (head) declines spread out. At this point, the pumping begins to capture water from springs, streams, plants, and other sources. The process increasingly shifts from reducing groundwater in storage to capturing water from other sources. If the pumping remains constant (and the stream hasn't completely dried up), the system will eventually reach a new equilibrium. At that point, the capture equals the pumping rate, which means water is no longer drawn from storage and water levels throughout the system are stable. The long-term source of water to a pumping well thus translates into reduced availability of surface water and less groundwater for plants.

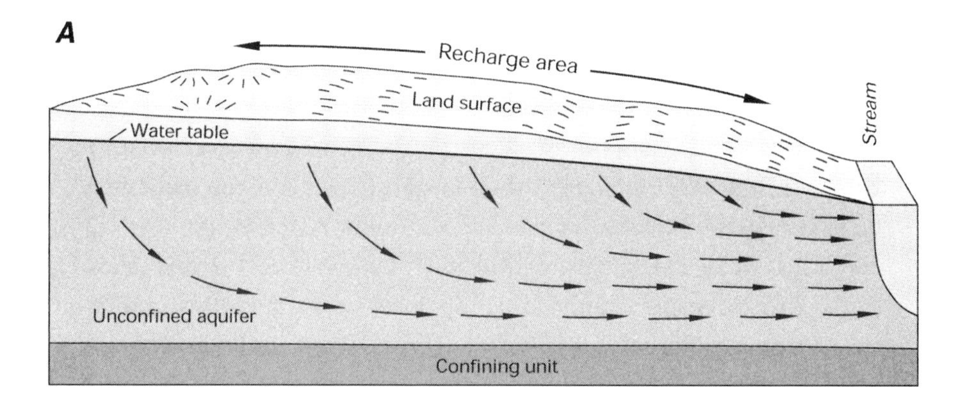

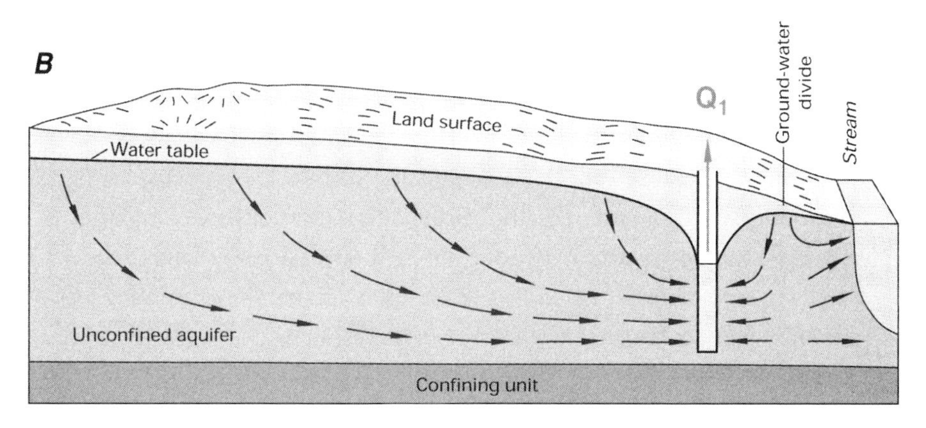

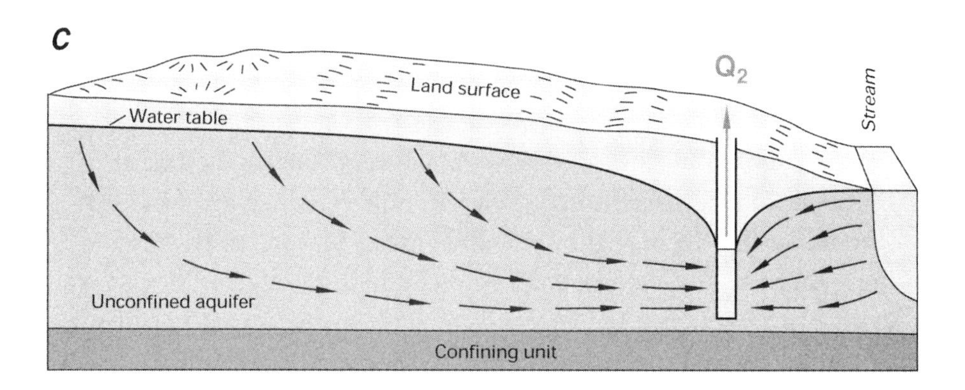

Effects of pumping from a simple unconfined aquifer: (A) Natural conditions. (B) Pumping intercepts part of the groundwater that would have discharged to the stream. (C) Increased pumping draws water from the stream to the well. *Source:* U.S. Geological Survey Circular 1139.

This brings us to a common misperception that a groundwater system is safe as long as pumping doesn't exceed the average rate of recharge. Known as the "water-budget myth," such thinking overlooks the key importance of groundwater to sustain surface-water flow and the environment. Pumping less than the rate of recharge can still cause adverse effects on these resources.

U.S. Geological Survey hydrologists Lenny Konikow and Stan Leake have extensively studied the importance of capture. By examining numerous groundwater modeling studies, they found that on average about 85 percent of the total water pumped came from capture—with only 15 percent from storage depletion. In other words, reductions in groundwater storage are just the tip of the iceberg. The effects on surface water are larger, but much less obvious.[9]

There is yet another major complicating factor—the effects of pumping on surface water can be spread out over years, decades, or even longer. As a result, streamflow depletion can continue long after a well is shut down, and may not even peak until decades later. As we shall see, this delayed reaction greatly complicates the regulation of groundwater and surface water as a single resource.

While states ignore or downplay interactions of groundwater and surface water within their borders, they don't hesitate to point out the connection to their neighboring upstream states. Actually, they do more than just point out the connection—they take them to court.

Most of the large rivers in the western United States are governed by compacts that allocate water between states and establish a process for resolving disputes. These compacts, which are provided for under the U.S. Constitution, allow states to sign "treaties" with one another. Such compacts are enforceable commitments, not gentlemen's agreements. If a state fails to abide by the agreed-on rules, the other state(s) can appeal to the U.S. Supreme Court as the final arbiter. At the time of their signing, most river compacts failed to address the effects of groundwater pumping on streamflow, which has made enforcement particularly messy. Colorado offers some instructive examples.

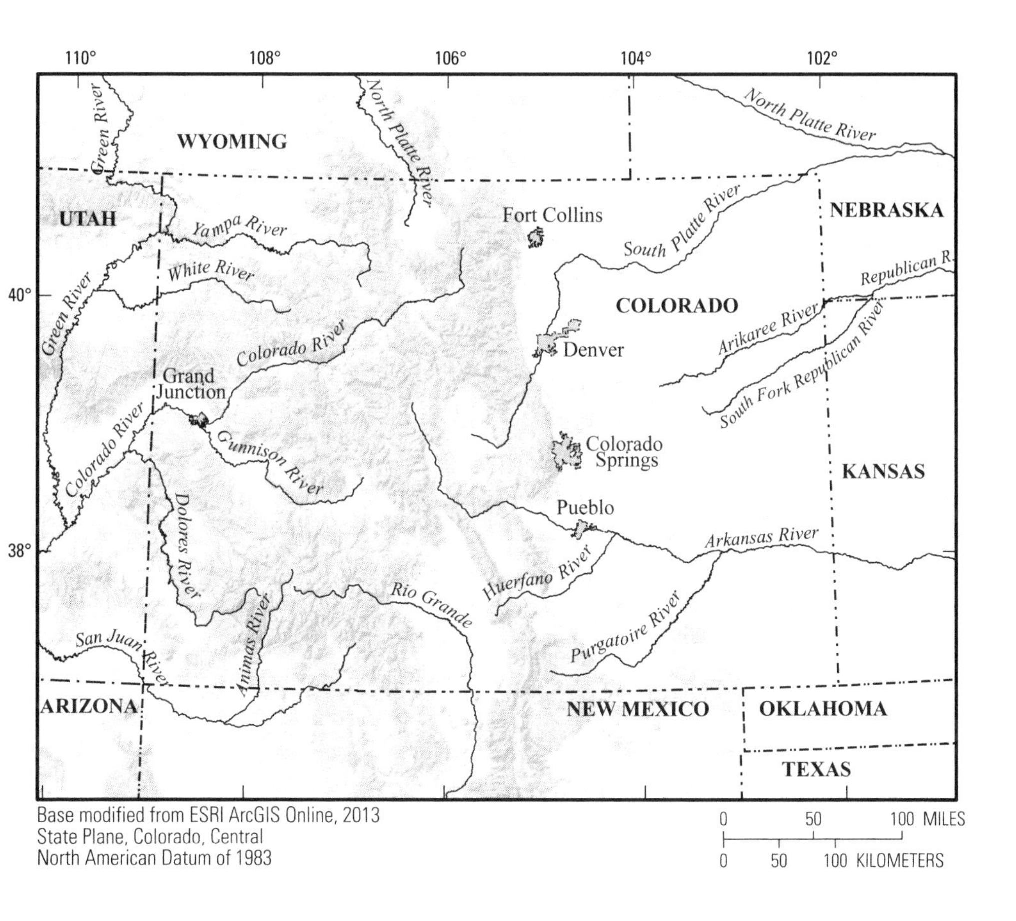

Major rivers in Colorado. *Source:* Michael Kohn, U.S. Geological Survey.

Straddling the Continental Divide, Colorado is a headwaters state with all of its major rivers originating internally. Three of these rivers—the Arkansas, the Rio Grande, and the South Platte—are overappropriated. The fourth, the Colorado River, is under extraordinary stress. As a result of nine river compacts, Colorado is legally bound to deliver about two-thirds of its streamflow to downstream states. With no upstream states to point fingers at, Colorado has been on the receiving end of several major U.S. Supreme Court cases.[10]

The opening salvo came in 1902 when Kansas sued Colorado in the Supreme Court, contending that Colorado was illegally consuming water from the Arkansas River that was owned by Kansas. If Kansas prevailed, Colorado would have had to pass the entire flow of the river to its downstream neighbor to the east. Colorado fought back, claiming that it owned all the water originating within its boundaries. Further complicating matters, the federal government joined the fray, claiming that the 1902 Reclamation Act reserved all remaining unappropriated water in the western states for future reclamation projects.[11]

The Supreme Court saw things differently, and no one prevailed. In its 1907 decision *Kansas v. Colorado,* the Court enunciated the "equitable apportionment doctrine," under which states (upstream and downstream) and the federal government must share the water equitably. As a further inducement, if the states couldn't agree on the allocation, the Supreme Court would determine everyone's fair share.

After nearly half a century of disputes over the Arkansas River, Colorado and Kansas finally signed a compact in 1948. Peace prevailed for a few decades. Then, in 1985, Kansas went back to the Supreme Court, alleging (among other things) that excessive groundwater pumping by Colorado farmers was depleting the flow of the Arkansas River, in violation of the compact. This time around, the Supreme Court ruled in Kansas's favor. Colorado was forced to make restitution, which amounted to about $35 million for undelivered water from 1950 to 1996. Colorado also had to adopt strict well metering and water accounting methods to assure future compliance.

The Republican River originates in the eastern High Plains of Colorado, briefly enters Kansas, then flows across southern Nebraska before reentering Kansas. In 1943, the three states ratified a compact on how to share the river. With no mountain headwaters, there's not much of a river to go around.

Groundwater pumping was minor at the time the compact was signed, but took off soon thereafter. Groundwater levels (storage) declined, but streamflow depletion (capture) became the bigger problem. By the 1990s, only 3 to 5 percent of the 350 million acre-feet (430 billion cubic meters) of groundwater in the Upper Republican River Basin had been depleted, yet stream baseflow had been cut in half.[12]

After years of dispute, in 1998 Kansas filed suit against Nebraska in the U.S. Supreme Court. The state alleged that Nebraska had violated the Republican River Compact as a result of overpumping groundwater. Nebraska countered that groundwater pumping was not governed by the compact, since the term "groundwater" was never mentioned. The special master, appointed by the Court to hear the lawsuit, denied Nebraska's motion to dismiss the case on the following grounds: "Although the Compact never uses the word 'ground water,' streamflow, which the Compact fully allocates, comes from both surface runoff and ground water discharge. Interception of either of those streamflow sources can cause a State to receive more than its Compact allocation and violate the Compact."[13]

Recognizing where the case was headed, Nebraska (and Colorado, which had been dragged into the lawsuit) agreed to a settlement in 2002. A moratorium was placed on new wells in much of the basin. Colorado already had a de facto moratorium in place, while Nebraska had continued drilling new wells. Under the terms of the settlement, both states needed to find ways to return water to the Republican River and send it downstream to Kansas. And they needed to do it quickly. Colorado's chief deputy state engineer, Ken Knox, compared this requirement to "President Kennedy asking NASA officials to make it to the moon on a deadline." Further complicating matters, 2002 marked the onset of a major multiyear drought.[14]

In 2004, the Colorado legislature created the Republican River Water Conservation District. Composed of local water users tasked to work toward compact compliance, the district instituted a fee on water consumed within Colorado's portion of the basin. The funds were used to pay farmers to voluntarily retire irrigated lands, but this still wasn't enough to get Colorado off the hook.[15]

In a controversial move in 2011, Colorado drained Bonny Reservoir near the Kansas border. The reservoir had long been an oasis for boating and fishing. Draining it helped Colorado meet compact compliance (credits included reservoir evaporation that no longer would take place). It also resulted in probably the biggest fish massacre ever connected to groundwater pumping. Fishermen caught as many fish as possible before the lake was drained, but the scene at the end was heart-wrenching. As the reservoir drained, workers with pitchforks removed fish that were clogging the outlet of the dam. Using a large mechanical crane, thousands of fish were scooped up and buried (many still alive) in trenches. There had been discussions of ways to euthanize them, but the state engineer had concluded that "digging trenches is the most humane way to bury the fish."[16]

Nebraska faced even greater challenges to meet its compact allocations with Kansas. With its abundant groundwater supplies, Nebraska had for many years been relatively complacent about regulating groundwater. In 1966, the Nebraska Supreme Court ruled that even where surface water and groundwater were well connected, they need not be viewed as such by law.[17]

The authority to administer and manage groundwater in Nebraska is granted to Natural Resources Districts governed by locally elected boards, which makes it challenging to come up with a comprehensive basin plan. In order to meet compact obligations, the state had to change its laws for groundwater regulation. It also had to figure out who would pay for compliance actions, and how they'd do so. Local water users resisted the state on these issues, slowing the process. A major breakthrough came in 2004 with passage of Legislative Bill 962, which officially acknowledged a hydrologic connection between groundwater and surface water and called for integrated management plans.[18]

Under the final settlement stipulation, 2006 was the first year when Nebraska's compliance with the terms of the settlement was determined. According to Kansas's calculations, Nebraska had used 79,140 acre-feet (98 million cubic meters) more water than it had been entitled to during the previous two years—tantamount to ten years of water use by a city with 100,000 residents. As the deficit continued to grow, Kansas did not have adequate water for users in parts of the basin.[19]

Colorado also had violated the compact, but entered into nonbinding arbitration to try to resolve the issue. This would be done in part by a $60 million, twelve-mile (nineteen-kilometer) long pipeline to pump water to the river from wells that formerly irrigated crops. Eventually, Nebraska also built two long pipelines to direct water from former irrigation wells to the river.[20]

In 2010, Kansas brought suit against Nebraska for continuing to violate the Republican River Compact agreement. This time the gloves were off. Kansas sought $72 million in damages, additional punitive damages, shutdown of wells irrigating 300,000 acres (120,000 hectares) of land, and preset damages for future violations. Preset damages meant that Kansas could calculate the amount of water owed in the future, multiply this by a dollar per acre-foot fine, and present the bill to Nebraska without the bother of returning to court. Worried that it was next, Colorado sided with Nebraska.

In February 2015, the U.S. Supreme Court ruled that Nebraska had knowingly violated the interstate compact governing the Republican River. "Nebraska recklessly gambled with Kansas's rights, consciously disregarding a substantial probability that its action would deprive Kansas of the water to which it was entitled," wrote Justice Elena Kagan for the majority. The court ruled that Nebraska owed Kansas $3.7 million (the value of the seventy thousand acre-feet of water) plus a $1.8 million penalty. Nebraska officials breathed a sigh of relief that the penalty was not larger. The court also amended the compact's water accounting procedures in favor of Nebraska.[21]

Despite the controversies in the Republican River Basin, it should be noted that in recent years Nebraska has become a leader in the integrated management of surface water and groundwater. The state has developed a unique water governance framework of twenty-three locally controlled Natural Resources Districts coordinated with a statewide Department of Natural Resources.[22]

Kansas is not the only state to sue over compact violations. In 1938, Colorado signed a compact with New Mexico and Texas on the Rio Grande. During the 1950s and 1960s, Colorado blatantly ignored the compact, sending water instead to farmers in its San Luis Valley to grow potatoes, barley, and alfalfa—and thereby accruing a water debt of nearly a million acre-feet (1.2 billion cubic meters). New Mexico and Texas sued. In 1968, the U.S. Supreme Court ordered Colorado to start meeting its annual compact obligations and to repay its past water debt. Saved by an act of God or nature, the water debt was cleared in 1985 due to a particularly wet year with high flows in the Rio Grande. Colorado continues to struggle to meet its year-to-year Rio Grande compact obligations.[23]

The 1923 South Platte River Compact was the second river compact signed among states (the Colorado River Compact signed in 1922 was the first). While the interstate disputes on the South Platte have been more amicably resolved, the river presents a classic example of the challenges of "conjunctive use"—the coordinated use of surface water and groundwater.

The South Platte River originates along the Colorado Front Range and flows some two hundred miles (320 kilometers) through Denver to the northeast corner of the state, where it continues into Nebraska. The alluvial aquifer connected to the river is among the most thoroughly studied and extensively used stream-aquifer systems in the United States. This is a long, narrow aquifer, two to six miles wide and up to two hundred feet (sixty meters) deep near the main river channel. Fertile soils in the basin produce irrigated corn, alfalfa, and vegetables. The corn and alfalfa support large cattle and dairy operations.[24]

Much of the irrigation along the South Platte is supplied by surface water delivered through irrigation ditches having water rights dating back to the late 1800s. As in all western states, surface water is allocated by the prior appropriation doctrine: first in time, first in right. By the 1890s, surface-water appropriations in the South Platte Basin exceeded the river flow during the irrigation season. Farmers unable to tap into surface water, or who owned very junior surface-water rights, increasingly turned to groundwater.

In the late 1960s, hydrogeologist John Bredehoeft and economist Robert Young teamed up to examine the potential benefits of conjunctive use of surface water and groundwater in the South Platte Basin. This was the first time that computer groundwater models had been used for economic analysis. Bredehoeft and Young demonstrated that conjunctive use in the South Platte Basin could double the net economic benefits compared to surface-water irrigation alone. Groundwater as a backup supply would also provide a form of drought insurance. Due to institutional and legal constraints, however, this potential of conjunctive use has not been realized.[25]

Colorado has been among the most progressive states in managing groundwater and surface water as a single resource. State law presumes that groundwater is tributary to streams and rivers unless proven otherwise, or given a pass by statute. Specifically, tributary groundwater is defined as any groundwater for which pumping would reduce the flow of a stream by more than 0.1 percent within a hundred years. This puts a lot of groundwater in the "tributary groundwater" category.

Colorado has three additional categories that regulate groundwater as largely separate from surface water. A special subcategory called (believe it or not) "not nontributary groundwater" straddles both camps. This subcategory, created for Denver's fast-growing southern suburbs, requires that a certain amount of the groundwater pumped be placed back into the stream to protect senior surface-water rights, but offers more flexibility than the strict tributary groundwater category.

There is no doubt in anyone's mind that groundwater in the South Platte Basin should be considered tributary groundwater in the state's classification system. The South Platte River acts as a drain for groundwater in the form of baseflow from the alluvial aquifer. Return flows from irrigation make a substantial contribution to stabilizing river flows. Prior to irrigation (and transmountain diversions from the Colorado River Basin to provide Denver and other Front Range cities with water), the river dried up or ran at a trickle in its lower reaches in the summer.

The connection between surface water and groundwater was mostly ignored until a severe drought in the early 1950s brought matters to a head. Many surface-water diversions were severely curtailed. Some farmers received no water for extended periods. While holders of surface-water rights were being short-changed, their neighbors with wells pumped all the groundwater they needed for irrigating their crops. More maddening was the growing recognition that all this pumping was depleting river flows. Yet the state lacked the authority to deny a new well or regulate existing wells. This became known as the "well problem."[26]

Under pressure, the state legislature passed various laws, but with little teeth. A 1957 law required a permit to drill a well, yet provided no basis to deny the permit for lack of water. Finally, in 1965, state regulatory authority over groundwater was codified with passage of the Colorado Ground Water Management Act. The act recognized the connection between surface water and groundwater, and handed responsibility for groundwater to the Office of the State Engineer—the same office that regulates surface water.

At first, the state engineer tried to regulate pumping based on the proximity of wells to a river. The assumption was that those wells nearest the river had the largest impacts—at least in the short term. This approach was challenged in court as unfair. Why should groundwater irrigators near the stream be the only pumpers regulated?

In 1969, Colorado brought tributary groundwater into the fold of the prior appropriation doctrine through the Water Right Determination and Administration Act. The act's goals were to maximize beneficial use of the state's water resources, and prevent injury to users with senior water rights. As would be seen, these two goals are not necessarily compatible within the rigid constraints of prior appropriation.

A priority date was assigned to each well. Invariably, these were very junior rights susceptible to being shut down whenever a "call" is made on the river. A call can be made by any senior water right holder whenever they are not receiving their full allotment. The call requests that the division engineer (the person in charge of water in the basin) shut down the upstream junior-water-right holders until the senior right is satisfied. To be able to respond to such requests, the division engineer tracks water flow along the river and at numerous river diversions.

For surface-water rights, responding to a call on the river is relatively straightforward. When the diversions are stopped, streamflow increases immediately. The situation is quite different for wells. A farmer may stop pumping, but the effects might not reach the river for years—long after the water is past due. Groundwater hydrologists have the expertise to estimate the timing of groundwater depletion, but it's virtually impossible to predict future calls on the river. A call on the river is far from a rare event. For most of the irrigation season, the South Platte serves only water rights pre-dating 1900. This means that in the absence of some sort of intervention, the priority system would result in widespread curtailment of wells during the irrigation season.

To allow a farmer to continue pumping after a call has been made, the 1969 act introduced the concept of "augmentation plans." The basic idea works as follows. In an alluvial aquifer where water is fully adjudicated, every well must have a plan of augmentation for replacing the streamflow depletion caused by pumping. This water must be supplied—in time, location, and amount—to prevent injury to senior surface-water-right holders. In effect, the well owner must convince the water court that they will be able to replace any "out of priority" depletions that affect the river. In this way, well owners can continue pumping while ensuring that senior surface-water rights are protected through replacement water when needed.

Replacement water usually comes in the form of leased or purchased surface water, or water recharged in ponds and pits during the nonirrigation season. Of course, finding surface water to lease or purchase when needed can be difficult in an overappropriated basin. Increasingly, farmers have turned to developing recharge basins to add water to the aquifer during the nonirrigation season. Finding water to recharge, however, also can be challenging—much of the flow in the South Platte during the nonirrigation season is needed to replenish surface reservoirs that were drained during the irrigation season. Farmers must also make the case that their recharged water will make it to the stream when a call on the river is in place.

To obtain approval of an augmentation plan, a farmer must submit an application (often prepared by an attorney) to the water court. Colorado is unique in having a well-established water court for each of the seven major river basins in the state, including the South Platte. The application must explain exactly where the water will be obtained, where the water is to be used and what it will be used for, how much water will be used, when and where augmentation water will be required, how much augmentation water will be needed, and how the augmentation plan will be operated. The application should be supported by an engineering analysis, usually prepared by a water resources engineer.

If all this isn't enough to make one pull up roots and move to the city, there's yet another considerable challenge. Because the exact timing and amounts of future shortages (calls) on the river are unknown, groundwater users must hedge their bets on the amount of future replacement water that will be needed. In other words, well owners have to operate as if a drought could occur next year, or for several years, depending on the court decree.[27]

Court approval of an augmentation plan takes years. To circumvent this problem, in 1974, the Colorado General Assembly passed legislation that allowed the state engineer to approve temporary augmentation plans pending court adjudication (approval) of the final plans. Three years later, the General Assembly did a rethink and repealed the authority. Nonetheless, the state engineer continued to approve temporary augmentation plans for many well owners on a year-to-year basis—while warning irrigators that this was only a temporary, stop-gap measure. This practice would precipitate a major crisis, but the day of reckoning was put off for two decades. From 1980 to 2000, the South Platte experienced some of the wettest years on record, masking the underlying water-supply problems. During this period, cities along the Colorado Front Range grew rapidly, increasing their own water needs.

The first shoe dropped in December 2001, when the Colorado Supreme Court ruled that the state engineer lacked the authority to approve temporary augmentation plans. Although water was a side issue in the case, the ruling raised a major red flag for the owners of thousands of wells operating on temporary plans. The Colorado legislature responded by allowing the state engineer to approve temporary augmentation plans until January 1, 2006, giving the well owners a brief period of amnesty.

The second shoe dropped in 2002, with the worst drought in Colorado's recorded history. Senior surface-water users kept a call on the river almost constantly for the next three years, while well owners found it increasingly difficult to obtain affordable water to meet their augmentation requirements. Under Colorado law, ownership of a surface-water right is viewed as a property right that can be bought and sold for whatever the market will bear. Cities along the Colorado Front Range were purchasing agricultural water rights in the South Platte River Basin, thereby driving up the cost of replacement water even under normal flow conditions. By May 2002 (just a few months into the drought), water lease prices skyrocketed from $30 to $600 per acre-foot. Augmentation plans that had worked for decades were becoming untenable.[28]

By May 2006, the drought had a stranglehold on the South Platte Basin. Reluctantly, the state engineer shut down 449 wells affecting thirty thousand acres (twelve thousand hectares) of farmland of the Central Well Augmentation Subdistrict (WAS)—an association of groundwater pumpers who had banded together for the purpose of obtaining replacement water. Having been given some earlier assurance that they'd be allowed to pump at least a 15 percent quota during the 2006 growing season, the farmers had already planted an estimated $1 million worth of crops. With a diminishing snowpack in the Rockies, deals for lease water fell through. Under state law, the state engineer had no recourse but to shut down the wells.[29]

The order came swiftly. All Central WAS members were ordered by certified mail to stop pumping immediately. Those who refused to accept certified mail found notices posted on their wells. Power and flow meter information was collected from almost all well sites to verify compliance with the stop-pumping order. Central's manager, Tom Cech, expressed the farmers' outrage: "It's a sad state of affairs when Colorado water law is used to put good people out of business, in the guise of protecting senior water rights."[30] All told, the state engineer shut down more than a thousand irrigation wells. The economic costs through 2007 were more than $28 million, but the social and emotional costs to affected farmers and farm communities were immeasurable. Many landowners who depended largely on wells went bankrupt.[31]

Today the South Platte River Basin has many fewer operating wells. In 2002, there were 8,200 permitted wells in the basin. In 2010, 4,500 of these wells were enrolled in augmentation plans and continued to pump, although most were partially curtailed. The remaining 3,700 wells had been completely shut down.[32]

The seemingly endless problems and conflicts over the South Platte then took an unexpected turn. As groundwater levels rose in response to the reduced pumping and increased recharge for augmentation plans, people in some communities along the river began to experience flooded basements, failed septic systems, and waterlogged fields. A representative from the governor's office stated the obvious: "How surreal it must be to have water in your basement, yet you can't turn on your well."[33]

The drought finally broke. Then in 2013, one of the worst floods in Colorado history inundated the mountain streams that feed the South Platte. Surface reservoirs filled and the need for groundwater pumping was temporarily reduced.

The story of the South Platte and other basins in Colorado is a familiar one for western stream-aquifer systems. In the beginning, groundwater is developed under minimal or no regulations. Sooner or later, it becomes obvious that groundwater pumping is reducing streamflow. The streamflow depletion deprives water from senior surface-water right holders, who seek redress from the courts. The courts uphold the priority rights of the senior users and agree that groundwater pumping is adversely impacting these rights. The state is forced to act in crisis mode. By this time, there are so many wells that it's much too late to simply restrict further groundwater development as a solution to the problem.[34]

Making use of the large volume of groundwater in storage through a conjunctive groundwater and surface-water system has the potential to greatly enhance the efficient use of water and help the world meet its growing food and other water demands. Approaches that lead to shutting down wells during droughts, however, clearly work against the maximum beneficial use of groundwater. Solutions to the "well problem" require more cooperative arrangements tuned to local conditions. A pivotal challenge is that the dynamics of groundwater are very different than those of surface water. Addressing these challenges is best done sooner rather than later.

It's not just western rivers where groundwater and surface-water interactions have come to a head. Extensive groundwater pumping has resulted in significant streamflow depletion in the Flint River in Georgia, further exacerbating the battle over water between Georgia and Florida in the ACF River Basin. In Wisconsin, the connection between groundwater and surface water is strongly contested, amid charges that irrigation wells in the central part of the state have caused streams and lakes to dry up.[35] The list of areas needing to address this challenge keeps growing.

8

Water for Nature

One day you may have to tell your grandchildren stories about places like this.
—Ad by The Nature Conservancy showing a lush deciduous forest

Theodore Roosevelt was the first president to speak out on conservation. When his wife, Alice, died two days after giving birth in February 1884, and his mother died the same day (in the same house), he was in despair. Roosevelt temporarily left politics to take up cattle ranching in the Dakotas, where he witnessed firsthand how human activities were harming the environment. Buffalo were being slaughtered on a vast scale, and the Transcontinental Railroad made it easy to transport their highly valued hides to market. In just two decades, the great bison that had once darkened the plains as far as the eye could see had been hunted nearly to extinction. When Roosevelt wrote about them in 1893, fewer than five hundred buffalo existed. Roosevelt recognized that without dramatic action, the country's rich natural resources and priceless landscapes would disappear as quickly as the buffalo.[1]

As president, Roosevelt set aside 150 national forests, fifty-one federal bird reservations, five national parks, eighteen national monuments, and four national game preserves. Every one of these designations was bitterly opposed by commercial interests. Roosevelt appointed Gifford Pinchot, who shared his belief in conservation through sustainable use, as the first chief of the U.S. Forest Service.

On May 13, 1908, Roosevelt's speech at the Conference on the Conservation of Natural Resources reflected his visionary thinking about the need to preserve the natural world around us:

> We have become great because of the lavish use of our resources and we have just reason to be proud of our growth. But the time has come to inquire seriously what will happen when our forests are gone, when the coal, the iron, the oil and the gas are exhausted, when the soils have been still further impoverished and washed into the streams, polluting the rivers, denuding the fields, and obstructing navigation. These questions do not relate only to the next generation or the next century. It is time for us now as a nation to exercise the same reasonable foresight in dealing with our great natural resources that would be shown by any prudent man in conserving and wisely using the property which contains the assurance of well-being for himself and his children.[2]

Today's national forests encompass a combined area larger than Texas. The Forest Service's mission involves much more than the task of overseeing this immense land area, and is complicated due to a number of competing and conflicting interests. When most people think of the national forests, they picture hiking through untrammeled wilderness, and kayaking or fishing on pristine mountain streams. This is part of what the national forests are about and, as such, the Forest Service must manage these public lands with an eye to protection and conservation. However, unbeknownst to most people, the Forest Service is part of the Department of Agriculture. Unlike the national parks, which were created to preserve natural beauty and unique outdoor recreational opportunities, the national forests were set aside as working forests with the two-pronged mission of responsible conservation and resource use. The Forest Service's original mandate was fairly simple and straightforward: to protect watersheds and furnish a continuous supply of timber. When it was discovered that the national forests contain some of the country's richest mineral deposits, it wasn't long before mining (including oil and gas production) entered the picture. The environmental consequences from these commercial activities means the Forest Service must deal with contamination and ecosystem destruction.

The Forest Service's job of protecting watersheds has become increasingly complicated. The original task basically involved controlling erosion from logging so that streams didn't fill with sediment. With population growth and development, particularly in the arid West, watershed protection has taken on new urgency. National forest lands are the largest single source of water in the United States, and form the headwaters and recharge areas of a large part of the nation's drinking-water supply. In the West, national forests provide proportionately more water because they include the major mountain ranges and headwaters of the principal rivers. More than nine hundred cities, such as Denver and Salt Lake City, rely on national forest watersheds for their public water supply.[3]

Protecting watersheds has become even more challenging as the role of groundwater in sustaining surface water and ecosystems has become increasingly evident. By the 1990s, hydrologists were increasingly adopting the view that groundwater and surface water are a single resource. During this period, Forest Service officials began to think about water in a much broader, and deeper, sense.

In May 2014, the Forest Service released its *Proposed Directive on Groundwater Resource Management.* The directive announced national guidelines by which the Forest Service proposed to manage the groundwater underlying national forest lands and protect groundwater-dependent ecosystems. Groundwater and surface water would be treated as a connected resource, unless it could be demonstrated otherwise. Something like guilty until proven innocent.[4]

The directive read like an innocuous and commonsense proposal to hikers, bird watchers, nature lovers, and outdoor enthusiasts. More than 125 groundwater scientists, legal experts, and conservation groups banded together and sent a letter to the Forest Service chief and the Secretary of Agriculture supporting the new directive and encouraging them to embrace their "role and duty to protect and sustainably manage water originating in and passing through National Forest lands."[5]

With all this laudatory support came the hardly unexpected resistance from state governments, to whom the Forest Service directive read like a declaration of war. During the public comment period, the water-strained western states objected in the strongest possible terms to the directive. In their view, the Forest Service's new "water management" role was in direct violation of the time-honored fact that only states have the legal mandate to manage and regulate water in their state, including on national forest land. Furthermore, the states have separate laws governing surface water and groundwater and were not going to be told that they had to prove anything to the Forest Service. The Western Governors' Association raised the hue and cry that the directive amounted to "superseding states' authority to issue water rights." They also called the single-resource viewpoint a "rebuttable presumption." Comments from state water managers (and their lawyers) were extensive, exhaustive, combative, and all fell within the same basic sentiment of telling the Forest Service to go take a hike.[6]

The battle over what rights the states and feds have concerning activities on federal lands has been going on, in one version or another, for over a century. States interpreted the Forest Service directive as yet another egregious demonstration of federal government overreach. With abundant legal precedent backing them up, the states insist that the law is on their side. The Forest Service's admittedly scanty legal arsenal does, however, have one hefty weapon, in the form of the special-use permit, which gives it broad discretionary powers to manage activities on national forest land.

In 1976, Congress passed the Federal Land Policy and Management Act (FLPMA). The act stipulates that "public lands be managed in a manner that will protect the quality of scientific, scenic, historical, ecological, environmental, air and atmospheric, water resource, and archeological values." FLPMA also states that "where appropriate, federal agencies have the right to preserve and protect certain public lands in their natural condition" in order to minimize damage to scenic values and fish and wildlife habitat. In addition, water cannot be diverted from, or be transported across, national forests unless the Forest Service issues a special-use permit.

FLPMA is the primary legal basis behind the special-use permit for wells and water diversions, which empowers Forest Service managers to decide who gets to do what on national forest lands. In the real world, these powers aren't realized without a certain amount of bureaucratic rough-and-tumble with competing interests, but it does make the Forest Service something of a heavyweight gatekeeper who must be appeased to get through the door. Obtaining this coveted permit for a well or pipeline involves jumping through time-consuming and expensive hoops, at the end of which "a permit for a well or pipeline may be denied if the agency's analysis indicates that NFS [National Forest System] resources, including water, will not be adequately protected."[7] Needless to say, the special-use permit drives state governments and mining companies up the wall.

So we've got the national forests, which everyone (for one reason or another) loves, and we've got water, which everyone (whether they realize it or not) loves even more. The issues involved are about as complicated as issues can get, but the basic question is elementary. How long is a national forest going to last if there isn't enough water to sustain that forest? It's the same problem as establishing a military base where there isn't enough water for the troops. To make sure those kind of doozies don't happen, when Congress designates federal land for a specific purpose, it also reserves sufficient water for that purpose. This is known as "federal reserved water rights"—or "reserved rights" for short. And just to make sure all the bases are covered, these reserved rights are "implied" rights, meaning that the right exists even if Congress forgets to mention it when designating the federal land. This all looks good on paper, and certainly qualifies as a simple, commonsense approach for dealing with the human propensity to overlook important stuff. But in the world of water nothing is simple, which means these reserved rights have a way of becoming highly controversial. And as far as common sense goes, reserved water rights didn't even exist before several hard-fought battles that resulted in U.S. Supreme Court decisions.

Among the Supreme Court's most pivotal cases, the 1908 decision in *Winters v. United States* established reserved rights for Indian reservations. The case arose from the Gros Ventre and Assiniboine tribes who diverted water from the Milk River in Montana to irrigate their reservation lands. Henry Winters and other homesteaders came along a few years later and built a series of upstream diversion dams that left scant water to meet the tribes' needs. The homesteaders argued that the federal government (and hence the tribes) had relinquished any rights to water from the Milk River when Montana was admitted to the Union. The Supreme Court disagreed, ruling that sufficient water had been "implicitly reserved" for tribal use when Native American reservations were established. The priority date for tribal water rights became the date that the reservation was established, thereby giving tribes very senior water rights. This decision is now widely known as the Winters Doctrine.

Decades later came the 1963 Supreme Court decision in *Arizona v. California,* the case that gave Arizona its hard-fought victory over California for Colorado River water. This landmark case also extended the concept of reserved rights to federal lands. Once again, the priority date would be when the federal lands in question were first established—often in the late nineteenth and early twentieth centuries—giving these rights seniority over many existing private water rights.

In 1976, the decision in *Cappaert v. United States* extended the reserved rights doctrine to groundwater use on or near federal land. This case involved Devils Hole, an open fissure in Death Valley National Park where a warm water pool about fifty feet (fifteen meters) below the land surface is home to a small population of pupfish—so named because they dart about and chase each other like playful puppies. The endangered Devils Hole pupfish descend from Pleistocene-age ancestral stock and are found nowhere else on Earth. In the late 1960s, the Cappaerts, whose ranch surrounds Devils Hole, drilled several wells for irrigation. Pumping from these wells lowered the water level in Devils Hole and threatened to expose the narrow, submerged shelf on which the pupfish feed and breed. In the ensuing court case brought by the National Park Service, the Cappaerts and the state of Nevada argued, among other things, that surface water and groundwater should be treated as distinct. Unconvinced, the U.S. Supreme Court ordered pumping to cease, declaring that the evidence showed that groundwater and surface water are "physically interrelated as integral parts of the hydrologic cycle" and therefore the Cappaerts were causing the water level in Devils Hole to drop by their heavy pumping. End of story.[8]

In the mid-1970s, the Forest Service was taking note of these decisions and began to seek reserved water rights to protect instream flows for fish and wildlife. The agency was soon butting heads with western state governments, which viewed prior appropriation as the only law of the land. Under this legal framework, human uses of water were those uses traditionally emphasized as beneficial. Of course, this doctrine of human supremacy disregards the fact that humans aren't going to reign supreme for long, if we don't protect the hand that feeds us. One could also argue that fish and trees and all those other creations of Mother Nature were technically "first in time" and therefore "first in line." Under the doctrine of prior appropriation, however, you have to be able to prove that claim by having completed the requisite paperwork. No proof, no water—unless someone goes to bat for them.

Forest Service efforts on nature's behalf took a definite turn for the worse when it applied for reserved rights in the Rio Mimbres and its tributaries to protect fish, wildlife, and recreation in the Gila National Forest. When the local irrigation company objected, the New Mexico courts ruled against the Forest Service. The decision was affirmed by the U.S. Supreme Court. The basic problem for the Forest Service was that the Organic Act of 1897, the original legislation enabling federal forests, had mentioned only two purposes—protecting water flows and generating timber. The act was silent on fish, wildlife, and recreation.[9]

The 1978 Mimbres decision was a major blow to the Forest Service for obtaining federal reserved water rights, but by this time it had FLPMA and the special-use permit. Since then, the Forest Service has continued to pursue federal reserved water rights, but has been stymied by the states and the courts. This is pretty much where things stood in May 2014, when the service released its *Proposed Directive on Groundwater Resource Management* and inadvertently declared war on the western states. A year later, the Forest Service backed down and withdrew the directive.

The Forest Service directive illustrates how difficult it is to incorporate environmental protections within existing water laws. An alternative approach is to reserve water for ecological protection from the outset—a concept pioneered by South Africa.

Water availability is among the many inequities of South African history. Prior to the arrival of the Dutch in 1652, the indigenous people were hunter-gatherers or pastoralists roaming for fresh pasture and water for their cattle. Land and water were communally owned and free for everyone to use.

Around 1685, the Dutch East India Company began granting land ownership and the associated riparian water rights to Dutch settlers. The indigenous people no longer had access to land and water for their livestock, or for hunting wild animals drawn to water sources. With their traditional way of life becoming a thing of the past, many were forced to work on Dutch East India Company farms.[10]

Beginning in 1805, British rule further exacerbated the inequities of water availability as the government went to great lengths to supply water for the exclusive benefit of "Whites." The rise of the National Party in 1948 brought apartheid and the creation of official "non-White areas" in South Africa. Regressive policies toward non-whites only got worse after independence in 1961, when the majority black population became increasingly cut off from access to water and proper sanitation. By the early 1990s, about a third of South Africans had no access to a basic water supply and more than half lived without proper sanitation.[11]

In 1994, with the first democratic elections in South African history, the new government began the enormous task of redressing past inequities, including water resources. Providing basic water and sanitation services to the majority of the population became a high priority of the new government. As noted by Kader Asmal, the first post-Apartheid minister of the Department of Water Affairs and Forestry, "Many historically disadvantaged rural populations would today be stable agrarian communities (rather than unproductive and desolate) if they had historically been afforded equitable access to land and water resources . . . Any meaningful developmental approach must in part redefine the inherited low expectations of the historically disadvantaged."[12]

A new constitution provided the chance to start over. After years of isolation as a pariah state, the country actively sought international advice. The resultant National Water Act of 1998 is regarded as one of the world's most progressive pieces of environmental legislation. South Africa became the first country in the world to adopt national water legislation that serves to transform society based on social and environmental justice.[13]

The goals of the National Water Act can be summarized by the succinct phrase, "Some for all, forever." *Some* recognizes the tradeoffs inherent in the process and the realization that no entity will get all the water it might desire. *For all* encapsulates the desire to redress past inequities. And *forever* addresses the goal of sustainability.[14]

The act recognized surface water and groundwater as a single resource. It embraced a return to "the commons" in which the country's water cannot be privately owned and is held in trust by the state. As the trustee, the government must ensure that water is managed in a sustainable and equitable manner for the benefit of all. The Department of Water Affairs and Forestry (now Water Affairs and Sanitation) regulates and licenses all water use, except for households and some other small-scale uses.

The Water Act specified two "reserves" that have priority over other uses. The first assures that basic human needs for water are met for all South Africans. The second is designed to protect aquatic ecosystems.

The reserve for basic human needs, officially defined as twenty-five liters (6.6 gallons) per day per person, is small compared to the overall water supply. The key challenges are providing access to water and protecting water quality. Many South Africans still lack access to good quality drinking water, yet progress has been made. Between 1994 and 2006, ten million South Africans gained access to safe water for the first time.[15]

The purpose of the ecological reserve is to preserve the quantity, quality, and reliability of water required to maintain the ecological functions on which humans depend. As such, it is framed for the benefit of people and not to preserve nature per se. For example, fisheries are maintained with the utilitarian goal to sustain the livelihood of those who depend on fishing. Implementing the policies of this legislation has been compared to "climbing an uncharted mountain." The interconnected and complex nature of ecological systems presents a serious challenge that must be addressed in small steps.[16]

A large part of the challenge has been socioeconomic. Determining an ecological reserve requires not only a scientific basis, but also tradeoffs with development. The original idea was that all segments of society would participate in the decision-making process. The reality is something quite different. South Africa's majority poor black population, which has no history of involvement in water negotiations and organization, has been marginalized time and again.[17]

A major goal of the Water Act was to divide the country into nineteen catchment management areas, based on major watersheds. However, no additional funding was provided to develop them. By 2012, only two had been established and were functional. A Second National Water Resource Strategy was issued in 2013. Its goal is to learn from the mistakes of the past and place greater emphasis on water for the poor.[18]

South Africa has undertaken a bold experiment in water governance to achieve a balance among economic development, social equity, and ecological protection. It is one of just a few countries that spend more on water and sanitation than on military budgets. After more than two decades of trial and error, the country continues to struggle with implementation of its National Water Act. As of this writing, an ongoing drought is severely testing the nation with municipal and rural water supplies running dry.[19]

Ecosystems that depend on groundwater for their existence are more pervasive than commonly assumed. They occur in groundwater-fed streams and lakes, wetlands, caves and karst systems, springs and seeps, and coastal areas with fresh groundwater inflow. Springs support more than 10 percent of the endangered species in the United States, making them some of the most biologically and culturally important ecosystems on Earth. Abe Springer from Northern Arizona

University, who has studied more than a thousand springs, calls them the "canary in the coal mine" for groundwater health. Other groundwater-dependent ecosystems, such as wetlands, are important as habitats for migratory birds or rare invertebrate species. Examples of major groundwater-dependent ecosystems already discussed include the Edwards Aquifer in Texas (where the aquifer itself forms a groundwater-dependent ecosystem), springs in Australia's Great Artesian Basin, and the wetlands of Las Tablas de Daimiel in Spain.[20]

Primitive desert people were probably the first to recognize groundwater-dependent ecosystems because of plants (phreatophytes) that led travelers to oases and watering holes. The term phreatophyte comes from two Greek words, for well and plant. Such a plant is literally a natural well that lifts water from below the water table. Early hydrogeologists discovered that phreatophytes can be used to provide rough estimates of groundwater depth and quality. The depth to the water table determines which plant species abound. If the water table falls too far, species dependent on shallower depths die. The salinity and overall quality of groundwater have similar effects.

Groundwater interactions with surface water directly affect the ecology of surface water by sustaining stream baseflow and moderating water-level fluctuations of lakes. Groundwater exchange with surface water may be important continuously, seasonally, or only during droughts. Considerable attention has been placed on the hyporheic zone—the water directly beneath a streambed that is a dynamic mixture of groundwater and surface water.

The temperature of shallow groundwater is very stable compared to surface water—approximately equal to the average annual air temperature. Freshwater fish, such as trout, often depend on areas of groundwater discharge that keep streams from freezing in winter. Conversely, groundwater discharge to streams during summer provides refuge from excessively warm temperatures.[21]

Riparian vegetation usually requires a shallow water table to maintain high moisture content in the root zone. Water-table decline caused by excessive pumping can result in deterioration of the riparian ecosystem and a subsequent loss of biodiversity. This problem has come to a head in the San Pedro River in Arizona.

In the spring and fall, the skies over much of North America are filled with billions of songbirds, shorebirds, raptors, waterfowl, and insects winging their way north, and then again south, along their ancient migratory pathways. These seasonal mass movements of wildlife have intrigued humans for thousands of years and have resulted in some very creative explanations. Aristotle proposed that migratory birds spent the winter in wetlands, stuck in the mud. Yet until recently the big questions of what triggers migration and how animals navigate hundreds or thousands of miles, often returning to the same stream, patch of forest, or beach, remained shrouded in mystery.[22]

Thanks to advanced tracking devices, the secrets of migration are beginning to be revealed. The pioneering ornithologist Martin Wikelski describes this flood of new information as, quite simply, "mind-boggling." Wikelski and his team collected the first data on individual migrating insects. After gluing tiny radio tags onto fourteen green darner dragonflies, they discovered that they act a lot like birds—or rather, as Wikelski points out, "birds act like the bugs, since insects are more than 100 million years older." Just like migrating birds, the dragonflies stop to rest and refuel, take advantage of tail winds, and hunker down during stormy weather.[23]

Along with this flood of fascinating information came a sobering reality: many migratory species are now in sharp decline, some are seriously endangered, and one of the major reasons is habitat destruction along migratory pathways. Robert Glennon, author of *Water Follies,* compares migration under these circumstances to trying to drive cross country without any gas stations and restaurants.[24]

The most important migratory route in North America is the Central Flyway, along which birds migrate from the Sierra Madre of Mexico to the mountainous regions of the western United States and points north. The main artery of this flyway is the San Pedro River in southeastern Arizona, where almost four hundred species of migrating birds rest in the river's forests and feast on the abundant insects and plants along the river. There are, in addition, around a hundred species of year-round residents. The San Pedro River, which begins in Sonora, Mexico, and empties into the Gila River 170 miles (280 kilometers) to the north, supports almost two-thirds of all bird species seen in North America.[25]

This area is so special that *Birder's Digest* named it the premier bird watching site in the United States. Three major bird conservancies have designated the San Pedro as a Globally Important Bird Area—the first designation of its kind in the Western Hemisphere. The Nature Conservancy placed the San Pedro River Basin on its list of "Last Great Places" in the Western Hemisphere. Capping off these designations, in 1988, Congress created the San Pedro Riparian National Conservation Area—the first of its kind in the nation. The conservation area encompasses almost 57,000 acres (23,000 hectares) along a forty-mile (sixty-four-kilometer) stretch of the river. At the same time, Congress created a federally reserved water right to go with it.

Despite all the fanfare and protection, the river is in growing jeopardy. The nearby city of Sierra Vista and neighboring military base of Fort Huachuca are completely dependent on groundwater, and their withdrawals have been endangering the San Pedro River and its groundwater-dependent ecosystem.

Fort Huachuca is an updated and expanded version of the original mud fort constructed after the Civil War to provide safe haven for American troops during the desperate war with Cochise, chief of the Chiricahua Apache. Cochise's brilliant military tactics shut down settlement and travel through this part of the southwest for two decades. Ending the war finally required a generous and unprecedented peace treaty allowing the tribe to retain all of its vast tribal lands and remain on them. (The treaty was broken after things settled down for a while.) Since then, Fort Huachuca has grown into a sizeable military base with major economic benefits for Cochise County and the city of Sierra Vista. The fort is the county's largest employer. Thousands of additional contractors live in the area, most in Sierra Vista. The vast economic trickle-down means that nearly every facet of the city's goods and services is directly or indirectly tied to the fort.[26]

Shortly after creation of the San Pedro Riparian National Conservation Area, Fort Huachuca found itself at the center of controversy because of its massive water consumption on this semi-arid desert. There are also two endangered species in the area, the Huachuca water umbel (that obviously needs water to grow), and the southwest willow flycatcher. The fort got the message and adopted an aggressive water conservation program. They installed low-use plumbing fixtures and began to reuse treated effluent for watering the parade field, golf course, and sports complex. They designed systems to divert treated effluent and storm runoff to the aquifer. And just to make sure all the bases were covered, the fort commander told everyone to stop watering their lawns. These efforts paid off. By 2001, in just twelve years, the fort had reduced its water consumption by almost 50 percent. Fort Huachuca is no longer the principal threat to the San Pedro River.[27]

Meanwhile, Sierra Vista *is*.

Though too far from Tucson to qualify as a commute, even in today's marathon efforts to get to work and back, the city nonetheless continues to grow. Much of the boom is explained by location. Sierra Vista, Spanish for "mountain view," is an understatement. The area is nearly surrounded by five mountain ranges, with the tallest peaks often snow-capped in winter. Residential neighborhoods are flanked with hotels and shopping centers. Sierra Vista is one of those places where you can get away from it all without having to leave anything behind.

In the late 1980s, Sierra Vista began to attract some unwelcome attention. When Congress established the San Pedro Riparian National Conservation Area, the county hired Thomas Maddock, a well-regarded hydrologist at the University of Arizona, to study the impact of groundwater pumping on the river. Maddock's resulting groundwater-flow model estimated that nearly 40 percent of the groundwater being pumped was capture water that would have otherwise discharged to the San Pedro. Upon learning of their impact on the river, Sierra

Vista officials reacted with a mix of denial and anger. They began looking for someone to prove Maddock wrong. A headline in the local newspaper read, "City Officials Declare War on Enviro 'Enemy.'" One city councilman went on record saying, "All right, there may be 500 species of wildlife found along the San Pedro. My response is, so what? What benefit do these animals have for humans? We are the ones who rule supreme, and if a plant or animal can't adapt to our needs, then it's too bad."[28]

This doesn't mean that most people living here don't care about the San Pedro River. They do. They like it. It's a great place to take the kids for Sunday afternoon picnics and a fun-filled splash in the water. It's just that people's water needs and wants take priority, like pretty much everywhere else.

By the 1990s, a conflict was brewing between various stakeholders over water use in the Sierra Vista subwatershed. Environmentalists were increasingly worried about the river. Most others were focused on economic growth and development. The Sierra Vista subwatershed had few (if any) limitations on drilling new wells, expanding agricultural acreage, or putting in large subdivisions.[29]

In 1999, Arizona began the Rural Watershed Initiative (RWI) as the state's main policy framework for groundwater resource management outside the five Active Management Areas. The oldest of Arizona's seventeen watershed partnerships is the Upper San Pedro Partnership (USPP), comprised of twenty-one local, state, and federal agencies and private organizations whose task is to address the water needs in the Sierra Vista subwatershed. Then Congress really put their feet to the fire with new legislation designed to achieve sustainable groundwater use by 2011, in order to preserve the San Pedro Riparian National Conservation Area and maintain the viability of Fort Huachuca. Considerable funding was tacked on to get the job done. There was no penalty if they didn't achieve the goal, but not meeting this mandated timeline could "negatively affect critical federal decisions, such as the future of Fort Huachuca."

The USPP tried to have productive groundwater-management discussions, but the whole issue was too emotional and complex, with too many cooks in the kitchen. Ecology, population growth, politics, economics, and land-use issues were all merging in the Sierra Vista subwatershed.

Holly Richter came to the San Pedro in 1999, after working on other rivers for The Nature Conservancy. Her background is in interdisciplinary modeling, working at the interface between hydrology and biology. Holly has a personality (and a laugh to go with it) that's big in every way—in her unbridled warmth and exuberance, in her love of talking and listening to people, and in her unabashed and unadorned caring about nature in a big, big way.[30]

The Nature Conservancy works on community-based conservation, with the goal of becoming a part of the community. Holly began by opening a dialogue with ranchers, realtors, and locals about the river and what was really going on. She talked to people who lived along the San Pedro River and had known it their whole lives. Some people told her that it was dying, some told her it was dead, and some told her it hadn't changed in a hundred years. "Everyone was all over the map," she told us. We had just sat down in The Nature Conservancy's Tucson office. Bright April sunshine poured through the window, offering a view of the spectacular Catalina Mountains in the distance. When we met her in the lobby, Holly explained that she doesn't get into the Tucson office very often, as evidenced by the pile of mail in her arms. Her job is to help save a river, so that's where she lives and works. Holly's passion for her work became obvious as she told the story about her part in trying to save the San Pedro.

"I quickly realized we weren't going to find a solution until we understood the problem. And the big question, to my mind, was to find out what the San Pedro River actually looks like in the pre-monsoon time in June, the driest time of the year." So she started the wet-dry mapping project.

The first of its kind in the country, the wet-dry mapping was designed to be a citizen science project—a way to bring people together and get them involved. Around thirty people signed up. The logistics were challenging and worrisome. People would be hiking through deep sand in temperatures of over a hundred degrees (thirty-eight degrees Celsius), mapping the wet and dry areas while keeping an eye out for snakes. Holly organized training events, borrowed GPS units, lined up shuttle vehicles, and made sure everyone had a first aid kit. Then she divided the forty-mile (sixty-four-kilometer) stretch of river into segments. Everyone got a piece.

Holly teamed up unlikely combinations of stakeholders ("Put it in the blender," she explained), pairing up an environmentalist with a city council member, a rancher with a developer. "It gave them a chance to get to know each other out in the wilds. And if you get to know someone, you're probably going to be a lot more polite and maybe even listen to what they have to say."

The day was a success—no snake bites, the first detailed map of where water starts and stops, and opposing stakeholders starting to work together as a cohesive team. "There was enough enthusiasm and interest to do it again the next year. And we're still doing it. We've done it every year since 1999, and it's grown to not only encompass the original 40 miles but the entire San Pedro, from where it starts in Mexico all the way to the Gila River and the tributaries up in the Huachuca Mountains." In June of 2015, no fewer than 143 volunteers from both countries mapped 302 miles (486 kilometers) of the river and its primary tributaries. Holly maps about twenty miles each year with Mike, a local rancher and retired math teacher from Tombstone. They ride their horses and map a section of the riverbed that is inaccessible by foot. "Mike with the reins in his teeth, writing down data," she laughed. "He's just amazing. He's a math teacher, right? So he's great for data. It's been a fascinating social engagement, as well as the fact that the information's been really useful. Now we can tell trends in any given reach, or the whole river over time."

The wet-dry mapping project has become the river version of an adopt-a-highway program. Holly explained how people have developed such a connection to their section of the river that they insist on mapping the same section year after year to see how "their place" is doing.

The now sixteen-year wet-dry mapping project has provided an invaluable picture of the San Pedro's hydrologic conditions over time, but it doesn't address what will happen in the future. As the cone of depression of water levels from pumping continues to move closer and closer to the San Pedro, it will increasingly jeopardize the river and the bordering woodlands. Such predictions require a calibrated groundwater model. The USPP recognized that the only way stakeholders could break out of their standoff was to have a groundwater model on which they could all agree.

This was no easy task. The partnership was created, at least in part, in response to "dueling hydrologists" (hired by different factions) who had provided very different results about how pumping was affecting the river. "You know how it is," Holly told us. "You can run a model and it can say all kinds of things according to what you're asking it, the parameters, the assumptions, the duration of the scenario. There are a hundred things, right? I mean, you can have the same exact model dueling with itself if you really wanted to. So the whole focus of the Partnership became to create a groundwater model that everybody could believe in and use for making decisions." They brought on board the U.S. Geological Survey (USGS). The job of modeling the Sierra Vista subwatershed was given to Don Pool and Jessie Dickinson, USGS hydrologists at the Tucson office.

The partnership also adopted an almost unheard-of strategy. Most scientific studies are presented in final form to the people who paid for it. Typically, most (if not all) of these people are nonscientists, which means they usually have a limited understanding of what the study shows. The result is predictable. Instead of science guiding the decision-making process, it soon gets muddled and lost in a tug-of-war among stakeholders. To avoid this problem, the USPP decided to be actively involved every step of the way.

Everyone agreed that the model needed to be very rigorous. They had long discussions about the holes in the knowledge, modeling assumptions about the hydrogeology, and what all the inputs to the model were going to be. The partnership worked closely with the USGS for seven years as they developed the model. Survey scientists were careful to make sure that all the members understood each component of the model by hosting question-and-answer sessions after presentations of each study. By the time the model was "built" everyone understood it, because they had understood all the parts along the way. As a result, they trusted it and used it for guiding decisions.[31]

An innovative use of the model was developed by Stan Leake, a research scientist at the USGS Tucson office, who created color-coded "capture maps." For any place on the map, and for any given model layer, these maps show what fraction of pumping will result in capture from the river in ten years, and in fifty years. With this invaluable tool, the partnership could see where pumping would most affect the river in the future. Leake then developed maps showing how recharge at any given spot will reduce capture. The model and capture maps gave the USPP the key to where and when to control pumping and start recharging.[32]

By using science as a focus for building mutual understanding, and agreeing to base decisions on scientific data, the USPP was able to find common ground and begin overcoming political differences to make decisions collaboratively. Initially, environmental issues were often misunderstood or disregarded. As their scientific savvy grew, special interests began to give way to decisions that were in everyone's best interest.[33]

The partnership took on a life of its own and evolved into a complex organizational structure. There were various committees—an advisory committee, a technical committee, an outreach committee, a government affairs committee, and so forth—all with great participation. Holly chaired the technical committee.

When the model was completed, Fort Huachuca hired a consultant, Laurel Lacher, who began running different scenarios. Based in part on her track record with Fort Huachuca, The Nature Conservancy decided to hire her as well. The first question Lacher "asked" the model was: Where are the best places to locate recharge sites along the river to sustain flows? "A form of river triage," Holly explained, "to stop the bleeding." Recharge along the San Pedro isn't going to eliminate the regional cone of depression, but it can help to mitigate its impacts on river flows. If everyone in the valley left today, the cone would still eventually impact the river.

They then asked Lacher to give them an idea of how much water it would take for recharge along the river to sustain flows at current levels for the next century—which came back to be a pretty reasonable number. "We don't have an ultimate solution," Holly explained, "but if we don't effectively manage the alluvial aquifer over the short term, and sustain it while we're figuring out the regional aquifer issue, we've lost it. I mean, how easy would it be to bring back this river if it becomes dewatered from an ecological standpoint? I don't even want to have to *ask* that question."

One of their best partners turned out to be the U.S. Department of Defense. Fort Huachuca is located in a unique setting because of its electromagnetic (EM) airspace, which allows the military to do communications testing that can't be done most anywhere else in the country. The high valley floor is surrounded by mountain ranges full of different metals that shield the electromagnetic airspace from cell phones, garage door openers, and all the other electronic gadgets now in use.

"This EM area is pristine," Holly explained. "If people build right around the fort, it messes everything up. So we said, let's create some buffers that preclude future development, don't allow for future pumping, are low-density residential, and keep the airspace open for you guys." The army can't buy private land and own it, so The Nature Conservancy agreed to acquire land and arrange for conservation easements with the private landowners. Under its Compatible Use Buffer Program, the U.S. Army invested $9 million in this effort.

"Then we used Laurel's different alternatives of the best recharge locations to sustain flows. We used the wet-dry maps to identify where flows have been interrupted. And we asked ourselves, what land is on the market that we could buy to establish a network of recharge facilities that would not only protect, but restore, the flows along the river? We ended up buying four properties in key locations with the $9 million. So now we have this beautiful palette of sites to work with, and we're in different phases for each site. One site, as of today, actually has graders. It looks like a new subdivision going in. It's actually a storm water facility testing infiltration trenches and dry wells, trying to understand storm water recharge on that site."

As the interview ended, Holly sat back with a look of calm satisfaction—looking forward, but possibly also looking back to those days when she was out beating the bushes, talking to people, and trying to figure out where to start. "The thing I've learned," Holly told us before we parted, "is how long it takes to get an outcome like this. The same people were at the table for a decade. And it was the people, not necessarily all of the institutions they represented, that made the difference in the long run."

9

That Sinking Feeling

Therein is the tragedy. Each man is locked into a system that compels him to increase his [use] without limit—in a world that is limited.

—Garrett Hardin

In 1939, Joe Poland was hired by the U.S. Geological Survey to study saltwater intrusion that was threatening public water supplies from Los Angeles to Orange County. Poland had just finished his master's degree at Stanford, with his Ph.D. at the "ABD" (all but dissertation) stage. One of his first classes had been Professor Tolman's course on groundwater geology, an entirely new field at the time. His main interest soon centered on land subsidence. During his years at Stanford, Poland worked with Tolman on the Santa Clara Valley (today's Silicon Valley), where significant land subsidence was occurring. After eight years at Stanford, Poland was impatient to get out and do some field-based research. He'd get back to the dissertation later.[1]

After being hired by the USGS, Poland studied saltwater intrusion in the Long Beach area. His first move was to round up as many water and oil drilling logs as possible—a formidable task. Vast oil reserves had been discovered in the Los Angeles Basin. By 1930, California was producing nearly a quarter of the world's oil. The Long Beach Oil Field included Signal Hill, one of the most productive fields on Earth. By the time Poland arrived in Long Beach, the entire area was a virtual forest of drilling rigs.

There was a lot more going on than drilling for oil. Long Beach was on its way to becoming the second busiest container port in the

United States, and the Navy was expanding its submarine base into a full-scale naval operations base.

Two years into his research, Poland became involved in a developing crisis in the harbor area. Naval installations were showing unexpected, and considerable, structural damage. Foundations of some dry docks and ship building facilities were collapsing. Poland's study of the area's oil well logs pointed to the newly recognized phenomenon of land subsidence induced by pumping oil.

When oil executives learned of this young scientist pointing the finger at their operations, they called him in for a meeting. Without mincing words, they informed Poland that they were men of considerable power and political influence. He could lose his job if he continued his investigation in the harbor area. A few weeks later, a recalcitrant Joe Poland was called to Washington to give expert testimony about the damage to the harbor area and its possible causes. From the time he left California, an FBI agent accompanied him around the clock until he completed his testimony. The oil companies were found to be responsible for the damage to the naval installations. Nonetheless, production continued and land subsidence at Long Beach eventually reached twenty-nine feet (8.8 meters)—about the height of a three-story building. The subsidence was eventually halted in the 1950s by pumping seawater into the oil field.[2]

In 1949, the USGS opened a regional office in Sacramento and put Poland in charge. Over the next few years, he supervised statewide investigations that led to the delineation of California's major aquifer systems and their storage capacity. In the mid-1950s, land subsidence from overpumping groundwater was becoming a widely recognized problem in the state. The worst-hit area was the San Joaquin Valley, which comprises the southern two-thirds of California's Central Valley and is one of the most productive agricultural regions on Earth.

Poland's work came at a fortuitous time. In the 1920s, Karl Terzaghi's investigations had led to the field of modern soil mechanics. Concurrently, Oscar Meinzer and other USGS scientists had pioneered the field of quantitative groundwater hydrology. The two sciences largely went their separate ways until investigations of land subsidence by Poland and others combined the two fields into the modern science of aquifer mechanics.

For the next two decades, Poland was project chief of a research program on the mechanics of aquifer systems, ultimately saving California millions of dollars through the redesign of freeway, irrigation, and aqueduct construction in subsidence-affected groundwater basins. One of Poland's legacies was his contribution to the California State Water Project, one of the world's largest water conveyance systems.

The State Water Project begins in the Sierra Nevada foothills ninety miles (150 kilometers) north of Sacramento, stretches south through the San Joaquin Valley and over the Tehachapi Mountains, and finally ends at a reservoir southeast of Los Angeles. The project consists of thirty-two dams and reservoirs; more than seven hundred miles (1,100 kilometers) of canals, pipelines, and tunnels; five power plants; and seventeen pumping stations. It's the biggest consumer of electricity in California. Moving water over the Tehachapi Mountains is a feat unto itself. The water is pumped two thousand feet (six hundred meters) over the mountains (initially via four mammoth pumps, each capable of powering a battleship), making it reportedly the highest single water lift in the world.[3]

The State Water Project was intended to supply a water lifeline to southern California cities and to relieve groundwater overdraft in the San Joaquin Valley, which was sinking at unprecedented rates. The proposed canal passed through areas with some of the most severe, ongoing subsidence. As the chairman of an interagency committee formed to study the problem, Poland developed the basic research strategies that guide subsidence studies to the present day.[4]

Much of Poland's working life focused on California, but for decades he was also an international leader on subsidence. In 1966, the highest tide ever recorded in Venice, Italy, completely flooded the island-city. UNESCO invited Poland to Venice to investigate why the city was sinking at such an alarming rate. Heralded upon his arrival as the "Healer of Venice" by the Italian newspaper *Il Tempo,* Poland subsequently promoted the scientific research needed to understand and control land subsidence in the city.[5]

In 1980, several of Poland's colleagues secretly arranged to have him reinstated as a grad student at Stanford, paid his tuition, and submitted a compilation of his most influential writings as a doctoral thesis. At the age of seventy-three, he finally received his Ph.D. Joe Poland continues to be remembered as Mr. Subsidence.

Land subsidence from overpumping groundwater in the San Joaquin Valley has been called the largest human alteration of the Earth's surface. When the last comprehensive surveys were made in 1970, subsidence in excess of one foot had occurred over more than 5,200 square miles (13,000 square kilometers) of irrigable land—half the entire valley. Southwest of Mendota, a town that prides itself on being the cantaloupe center of the world, maximum subsidence was estimated at twenty-eight feet (8.5 meters).[6] By this time, however, massive infusions of surface water were being delivered to the valley, and subsidence was slowing or had been "arrested."

Then came a series of droughts and cutbacks in imported water that resulted in renewed overpumping and subsidence. During the severe drought of 1976–1977, surface water imports were sharply reduced. The six-year drought beginning in 1987 was the state's first extended dry period since the 1920s into the 1930s. A severe drought from 2007–2009 marked the first time a statewide "proclamation of emergency" was issued. And then, in 2012, the worst drought on record began to grip the state. Paleoclimate investigations suggest that this was the most severe drought in 1,200 years, predating the Viking conquests in Europe.[7]

This drought wasn't just about lack of rainfall. What made it so extraordinary was the extreme heat that came with it. It was called the "Hot Drought." What this meant is that all those vegetables and orchards needed more water than ever. Some areas of the San Joaquin Valley were sinking by almost a foot (0.3 meters) a year.[8]

When groundwater is pumped from an aquifer system, hydraulic pressure decreases. This reduced pressure shifts the support for the weight of the overlying landmass from the water in the pores to the granular skeleton of the aquifer system. If the geologic materials

Approximate location of maximum subsidence in the United States identified by Joseph Poland (pictured). Signs on pole show location of land surface in 1925, 1955, and 1977. The pole is southwest of Mendota, California. *Source:* U.S. Geological Survey Circular 1182.

are sediments, rather than hard rock, the increased load on the sediments causes them to compact, with associated land subsidence.

Basic geology explains why subsidence affects the San Joaquin Valley and not the High Plains. The High Plains is basically a huge erosional sand pile from the Rocky Mountains. Anyone who has walked along the saturated tide-line of a beach knows that your footsteps quickly rebound and disappear. In the San Joaquin Valley, however, deposits of silt and clay (known as "aquitards") are sandwiched between, and within, aquifers. These clays are not only compressible, but if groundwater levels fall below critical thresholds, the compaction is mostly nonrecoverable—even if groundwater levels later recover. In other words, much of the land subsidence is permanent.

But that's only the beginning of the San Joaquin Valley's problems. Even if pumping returns to more normal rates, the subsidence will continue (albeit at a slower rate) long after water levels recover because of the slow drainage and response time of the aquitards. It will take decades for most of the pressure equilibration to occur, and for the ultimate compaction to be realized in some of the thicker aquifers in the valley. Meanwhile, things are never going to go back to the good old days. In the same way that a crushed soda can holds less water, nonrecoverable compaction leads to a permanent loss of aquifer storage. During the 1976–1977 drought, after only a third of the peak annual pumping volumes of the 1960s had been produced, groundwater levels rapidly declined more than 150 feet (forty-five meters) over a large area and subsidence resumed.[9] That a relatively small amount of pumping caused such a rapid decline in water levels reflects the reduced groundwater storage capacity caused by compaction. And that was four major droughts ago.

There had been no appreciable subsidence monitoring program in California since the last comprehensive survey in 1970.[10] The problem simply dropped off the radar. By the turn of the millennium, land subsidence in the San Joaquin Valley had disappeared as a major issue. The consequences of this oversight began to be realized during the 2007–2009 drought. The California Department of Water Resources conducted GPS surveys, two years apart, in an area near the critical

Delta-Mendota Canal. The remarkable difference in elevations sounded the alarm. Michelle Sneed, a USGS hydrologist who is an expert on land subsidence in California, was tasked to confirm the results. "Not only did we confirm the results," Michelle says, "but we found this very large subsidence area that was covering 1,200 square miles," an area about the size of Rhode Island.[11] Much of this area was away from historical centers of subsidence.

To help offset the monitoring gap, an innovative remote-sensing technology known as Interferometric Synthetic Aperture Radar (InSAR) allows scientists to measure changes on the Earth's surface as small as a few millimeters. A sensor on an Earth-orbiting satellite bounces radar signals off the ground surface. During repeat passes of the satellite over the same targeted ground surface, it is possible to precisely estimate changes in distances between the satellite sensor and the ground surface as it uplifts or subsides. The problem is that you have to know where to look. Surprises, like the new area near the Delta-Mendota Canal, had not been considered.[12]

During 2012 to 2015, instead of rain, the dominos were falling. A severely reduced snowpack in the Sierras lessened the streamflow, which led to inadequate reservoir supplies, which in turn reduced water allocations for much of the state. The situation was exacerbated by court-mandated reductions in surface-water deliveries in order to maintain adequate freshwater for fish and other environmental needs in the Sacramento–San Joaquin Delta. Incrementally, the State Water Project and the federally run Central Valley Project were providing less and less water. In early 2014, the drought had become so severe that both projects announced that they would cut off water deliveries to farmers in the San Joaquin Valley.

To save their crops, farmers began pumping at record-breaking rates from an already depleted aquifer system. Drillers had more business than they could handle deepening or drilling new wells. In some areas, groundwater levels dropped to more than a hundred feet (thirty meters) below previous historical lows. "I've been studying subsidence throughout the west for twenty years, and I've never measured rates like this before," noted Michelle Sneed.[13]

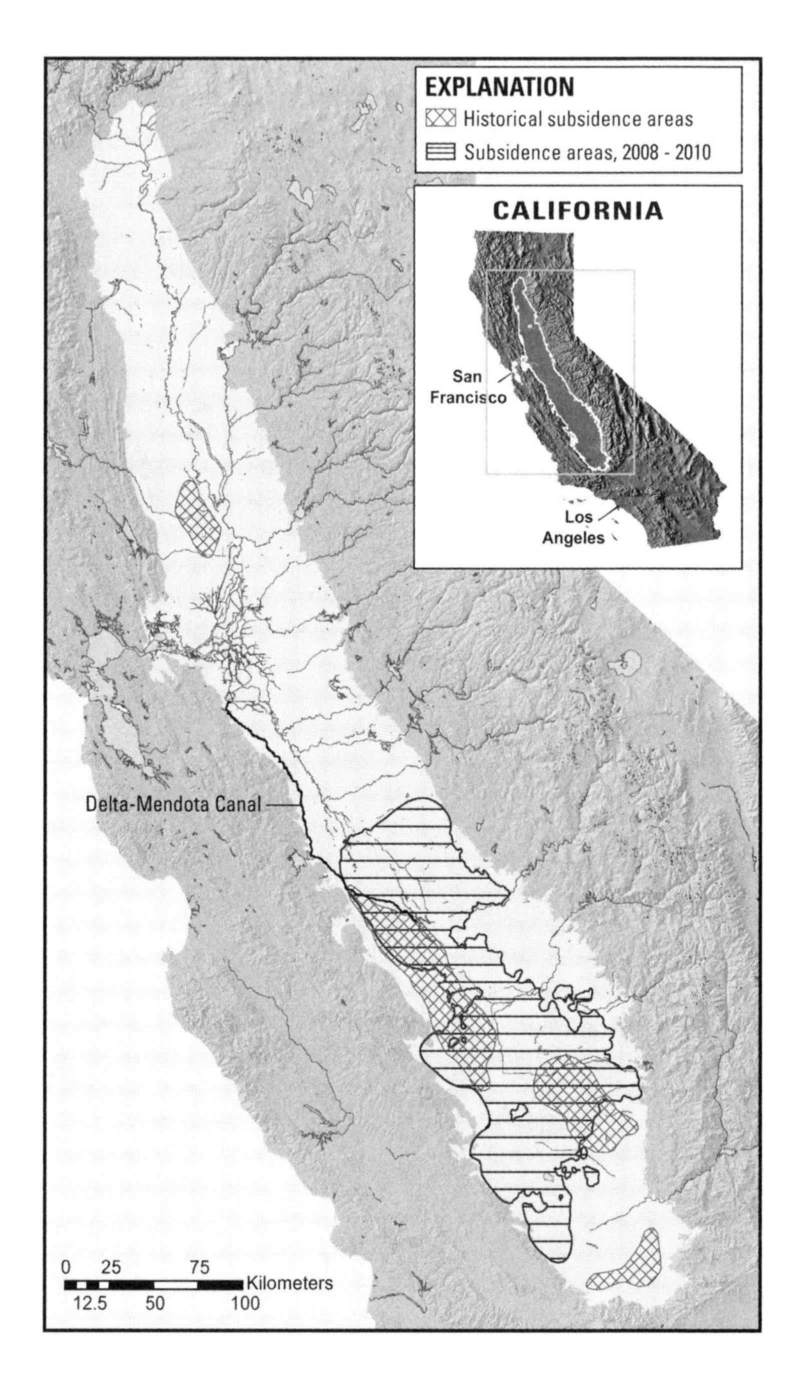

Historical and recently measured areas of land subsidence in California's Central Valley. Based on data from Claudia Faunt and Michelle Sneed, U.S. Geological Survey.

Infrastructure can't handle this unprecedented rate of drop. If the entire San Joaquin Valley were subsiding at the same rate and in the same way, it would still be bad—just not as bad. The problem is that different amounts of subsidence in different places are wreaking havoc with canals, pipelines, dams, levees, roads, railways, bridges, building foundations, sewer lines, and laser-leveled fields. Wells have also been damaged, because compacting clay causes well casings to buckle and eventually collapse.

Canals are particularly sensitive because subsidence affects the gradient that moves water by gravity in much of the system. When one part of a canal subsides, it reduces (or destroys) the conveyance capacity for all downstream parts of that canal. This makes downstream farmers even more dependent on groundwater. In addition, as the land sinks, bridges sink with it. One of the most extreme cases is the Russell Avenue Bridge north of Mendota. Before subsidence, canal inspections were conducted from a boat that passed easily under the bridge. Today, canal water laps against the bridge, thereby reducing flow capacity by about 45 percent.[14]

Bordered on both sides by mountain ranges, the San Joaquin Valley has a spectacular history of flooding. During a killer drought, this version of water abundance can almost sound good—until it happens. Both the Central Valley Project and the State Water Project were constructed with the dual purpose of supplying surface water and mitigating the frequent and devastating spring floods that can overwhelm the valley. The Eastside Bypass, a key flood control channel, is so severely impacted by subsidence that a large part of the valley is now in danger of massive flooding. At the Chowchilla Bypass, subsidence is so severe that its capacity to carry water away during a flood is expected to decrease by more than 50 percent. If the bypass is breached, an area a few miles wide and about twenty-five miles (forty kilometers) long could be inundated by as much as thirty feet (nine meters) of floodwater.[15]

Farmers in the San Joaquin Valley have developed an agricultural system dependent on plentiful imported surface water. This water wealth has allowed farmers to convert their fields from row crops to higher value nut trees (primarily almonds) and other permanent crops that require year-round watering. These vast fields can no longer be fallowed. When a drought hits and imported water is reduced, farmers start overpumping and subsidence continues. Michael Campana, an internationally recognized groundwater expert, sums up the problem: "Here we are 40-some years later, and they're still doing the same thing. It's like the classic definition of insanity."[16]

The Santa Clara Valley is the first area in the United States where land subsidence due to groundwater pumping was recognized. This is also the first place where remedial action was undertaken to effectively halt subsidence.

The Santa Clara Valley is a sediment-filled southward extension of San Francisco Bay. The Coast Ranges to the west are covered with lush forests and towering redwoods. Grass-covered hills form the eastern boundary. To early settlers, there seemed to be more water than anyone could ever use. Winter and spring rains filled ponds and rivers, bringing teeming flocks of wild birds and waterfowl. In addition, sand and gravel deposited by runoff from the surrounding mountains created good aquifers. The aquifers are recharged at higher elevations and confined by clay layers in much of the valley—conditions ready-made for flowing artesian wells. The first artesian well was dug in 1854 in downtown San Jose, then a small town of about three thousand people. By 1865, there were five hundred flowing artesian wells in the valley. By about 1915, as artesian pressure declined, windmills and pumps began to take their place.[17]

Groundwater was initially used for household needs and range-fed cattle. Farming was mostly dryland wheat. This simple, pastoral society began to change when railroads and refrigerator cars gave farmers access to expanded markets. Orchards, vineyards, and vegetable crops began to fill this "Valley of Heart's Delight." By 1920, two-thirds of the area was under irrigation, including 90 percent of the orchards. The Santa Clara Valley became the largest fruit growing and canning center in the world. Meanwhile, groundwater levels were falling.

In 1921, an engineering study initiated by farmers and business leaders recommended an ambitious project to construct seventeen large reservoirs to capture rainfall and replenish the aquifer through recharge. Initially, people were reluctant to invest in such expensive water projects. It was easier to believe that a few wet years would make the problem go away. When it didn't, the residents saw the writing on the wall. After a couple of failed attempts at the ballot box, the Santa Clara Valley Water District was approved on November 5, 1929. The approval was remarkable in that it came on the heels of the 1929 stock market crash.

Establishing the new water district and obtaining funds to construct recharge reservoirs were crucial steps, but unforeseen events complicated matters. World War II brought defense-related industry and burgeoning urbanization. The valley rapidly transformed into "Silicon Valley," the birthplace of the global electronics industry. Explosive postwar growth severely strained water resources and groundwater levels continued dropping. By 1964, the water level in a once-flowing artesian monitoring well in downtown San Jose had fallen to 235 feet (seventy meters) below the land surface.[18]

As water levels plummeted, sections of the valley floor began to sink. Subsidence was first noted in 1933, when benchmarks installed in San Jose in 1912 were resurveyed. More than three feet (one meter) of subsidence had occurred. In 1935 and 1936, the Santa Clara Valley Water District built five storage dams on local streams to capture storm flows. The dams allowed for controlled releases that would increase groundwater recharge through the streambeds. After a brief period of slowdown, subsidence resumed at an accelerated rate between 1950 and 1963. Then, in 1965, increased imports of surface water from the State Water Project allowed the Santa Clara Valley Water District to greatly expand its groundwater recharge program. Groundwater levels subsequently recovered, and little additional subsidence has occurred since the late 1960s.[19]

In some places, however, the damage was done. Land adjacent to the southern end of San Francisco Bay had sunk anywhere from two to eight feet, and the subsidence is permanent. As a result, the southern end of the Bay is ringed with dikes to protect the Santa Clara Valley from coastal flooding.[20]

Today's Santa Clara Valley has a well-coordinated program for the conjunctive use of surface water and groundwater. More than ninety miles (144 kilometers) of local creeks, and more than three hundred acres (120 hectares) of ponds, are used to replenish groundwater.[21] Most of the credit goes to the Santa Clara Valley Water District in addressing both subsidence and flood control.

Galveston, Texas, lies on a barrier island that separates the Gulf of Mexico from Galveston Bay and from Houston, the fourth largest U.S. city. In 1900, Galveston was a boom town. As a major deep-water port, the city had such a vibrant economy that it was known as the "Wall Street of the Southwest." Everything changed on September 8, 1900, when a surprise hurricane with winds clocked at 145 miles (233 kilometers) an hour slammed into this low-lying island. The next day, Galveston was a pile of rubble and eight thousand people were dead. Some estimates go as high as twelve thousand fatalities.

It was the deadliest hurricane in U.S. history. It was also the deadliest natural disaster ever to strike the United States. This single hurricane killed more Americans than the legendary Johnstown Floods, the San Francisco Earthquake, the 1938 New England Hurricane, and the Great Chicago Fire *combined*. The only thing this monster lacked was a name—that tradition began much later. Today's locals refer to it simply as the Great Storm.[22]

Possibly more than any other metropolitan area in the United States, Houston has been adversely affected by land subsidence. Houston, a flat low-lying city on the edge of Galveston Bay, receives nearly fifty inches (130 centimeters) of rain a year. Numerous local rivers are a constant threat to flooding. The nearby Texas Gulf Coast is subject to a hurricane or tropical storm about once every two years. Land subsidence in Houston is the last thing residents need—and they've got it in spades.[23]

When investors moved to Houston after the Great Storm, the area experienced rapid growth. A few years later the first successful oil well was drilled, marking the beginning of the massive petrochemical industry for which the city is known. In 1925, the U.S. Army Corps of

Engineers finished dredging a ship channel across Galveston Bay. Houston became the second largest deep-water port in the nation, and the eighth largest in the world.

Until 1942, essentially all of Houston's water supply came from groundwater. Subsidence had begun to affect a large area, although it was generally less than a foot. By the 1970s, however, pumping along the city's coastal areas had resulted in ten feet (three meters) of subsidence in places, with almost 3,200 square miles (8,300 square kilometers) having subsided more than one foot. The problem had reached crisis proportions. In 1975, the Texas legislature created the Harris-Galveston Coastal Subsidence District—the first of its kind in the United States. The district was authorized to issue (or refuse) well permits and to promote water conservation and education. The city also undertook conversion from groundwater to surface-water supply—a massively expensive undertaking. The district generally succeeded in arresting subsidence in the coastal lowlands, though subsidence continues in growing areas north and west of Houston.[24]

Subsidence in the Houston metropolitan area has increased the frequency and severity of flooding, causing extensive damage to industrial and transportation infrastructure. Major investments have been required in constructing levees, reservoirs, and surface-water distribution facilities. In addition, Galveston Bay is one of the most important and productive bay ecosystems in the United States. As a result of subsidence, more than 26,000 acres (10,500 hectares) of wetlands have been converted to open water and barren flats.[25]

The irony of Houston is that it's sitting on top of vast groundwater reserves that it can't use. The city's continued growth means that subsidence must be vigilantly monitored and managed. Today, however, the region is better positioned to deal with future problems because of raised public awareness, well-established subsidence districts with regulatory authority, and ongoing monitoring and research.

Surrounded by mountains, Mexico City lies in a closed basin that was filled by a series of connected lakes until the Spanish conquest. Over time, the lakes were drained by the Spanish overlords and the city was built on the highly compressible silts and clays of former wetlands and lake bottoms. Mexico City is one of the world's largest megacities, with a population of over twenty-one million, and is dependent on groundwater for more than 70 percent of its water supply. The combination of location and population has resulted in one of the largest subsidence rates ever measured. During the twentieth century, parts of the city sank as much as thirty feet (nine meters). Current rates of land subsidence can exceed one foot (thirty-five centimeters) per year. The cumulative damage to buildings, homes, and the city's infrastructure is thought to be equivalent to that caused by a strong earthquake.[26]

In the 1950s and 1960s, pumping was stopped in the city center after subsidence had damaged hundreds of treasured colonial churches and mansions. Since then, subsidence rates in the city center have slowed dramatically, but the effects continue. The Metropolitan Cathedral (the largest church in the Americas) leans to the left, requiring scaffolding to support the ceiling and walls, as well as a complex effort to shore up its foundations. The National Palace has received similar attention in an effort to keep one of its wings attached to the rest of the building.[27]

"The sinking of the soil in Mexico City is one of the biggest engineering problems any city has faced, anywhere," said Ismael Herrera Revilla, a mathematics professor at the National Autonomous University who led a five-year binational study of the city's water crisis in the 1990s.[28] The original sewer system no longer flows by gravity, requiring the installation of vast pumping stations and construction of more than a hundred miles (160 kilometers) of deep sewers to carry waste out of the city. Repairing ruptured tunnels in the city's metro system is an ongoing challenge. An aboveground line, constructed in the mid-1960s, began to look like a roller coaster.

Mexico City has one of the world's leakiest water distribution systems; it loses anywhere from 30 to 40 percent of its freshwater through leaking and ruptured pipes. It's a vicious cycle—continued population growth means more water needs to be pumped, resulting in more subsidence and more underground ruptured water pipes, which leads to even more water having to be pumped. The city has developed some costly alternative water sources by piping in water from considerable distances—but at enormous cost to those villages losing their water supply.

Mexico City's location in a seismically active region is another concern. In the eastern part of the city, where many of the public supply wells are located and maximum subsidence rates are occurring, reactivated faults and fractures can provide preferential flow paths for contaminants into the aquifer. The fissures are as long as several kilometers and 130 feet (forty meters) deep. Residents in this part of the city have become anxious about the possibility of a sudden fissure damaging their home, or worse. In 2007, a car and the body of its driver were found inside an open fissure at depths of fifty and seventy feet (sixteen and twenty-two meters), respectively.[29]

The effects of subsidence are most dramatic in Mexico City, but subsidence is also a problem in other cities in central Mexico.

Many large Asian cities are situated on deltaic flood plains and are vulnerable to land subsidence. Tokyo and Osaka (Japan), Jakarta (Indonesia), Ho Chi Minh City and Hanoi (Vietnam), Shanghai and Tianjin (China), and Bangkok (Thailand) are just a few of the cities where intensive pumping has caused land subsidence.[30]

Bangkok is located on the deltaic flood plain of the Chao Phraya River, the second largest delta in Southeast Asia. The Bangkok Basin, which lies beneath the southern half of the delta, consists of eight confined sand and gravel aquifers with intervening clay layers, and capped by the Bangkok Clay. Extensive use of groundwater in metropolitan Bangkok began in the mid-1950s, primarily for public water supply. Pumping for industrial uses followed. By the 1970s, pronounced land subsidence (and saltwater intrusion) began to occur.

Damages to urban infrastructure and increased flood risk during tidal surges spurred a transition of the public water supply from groundwater to surface water. Meanwhile, private use escalated and additional measures were needed. These included designation of "critical areas" where new water-well drilling was banned, licensing and metering of groundwater use, a groundwater use charge, and a special "groundwater preservation charge" for users in critical areas. The early establishment of a monitoring program to collect data on groundwater levels, land subsidence, and water quality has turned out to be invaluable for tracking progress. Fortuitously, the thick Bangkok Clay overlaying the system has limited the number of water wells drilled, and thus the number of users who have to be managed. Not everyone complies with the rules and regulating the private sector has been difficult; nonetheless, groundwater pumping peaked around 2000. Subsidence has been controlled in central Bangkok, but it continues to be a problem in outlying areas.

Land subsidence is one of the most damaging consequences of overpumping groundwater. The Santa Clara Valley, Houston, and Bangkok, as well as Tokyo and Osaka, have brought land subsidence largely under control. Many other places around the world, such as the San Joaquin Valley, Mexico City, Jakarta, and the Mekong Delta of Cambodia and Vietnam, continue to struggle with the problem.[31]

10

Recharge and Recycling

Water should be judged not by its history, but by its quality.
—Lucas van Vuuren

Centuries ago, tribes in the Karakum Desert of Turkmenistan dug trenches to drain rainwater to sand dunes underlain by clay.[1] Later, when supplies were short, the tribes would excavate the dunes to access the stored water. These nomadic tribes were practicing one of many ways that humans have devised to help nature replenish groundwater—a practice known today as "managed aquifer recharge." The previous discussions of this practice in India, Arizona, and the Santa Clara Valley are but a few examples of this growing solution to augment groundwater supplies.

Modern spreading basins and injection wells replenish aquifers with available water when there are surpluses, and recover the water when it is needed most. Recharge projects are also used to stop saltwater intrusion, halt land subsidence, or improve water quality by filtering water through the subsurface—a process called "soil aquifer treatment."

Underground water storage has some distinct advantages over traditional surface reservoirs. Many of the best sites for dams are already taken, construction costs of surface reservoirs are high, and considerable controversy often surrounds the environmental effects of tampering with the natural flow of rivers. Probably the most compelling argument is that storing water underground avoids massive losses to evaporation, particularly in hot, dry climates. One such example is

Lake Mead on the Colorado River. In a recent study, the U.S. Geological Survey estimated that about 6.5 feet (two meters) of water evaporates from the surface of the reservoir each year. Looking at it another way, every year an amount equal to almost twice Nevada's annual entitlement to Colorado River water simply vanishes into thin air.[2]

Underground storage comes with its own set of challenges. First and foremost, a suitable aquifer is required to store the water, which means you have to have the right geology. In addition, the recharged water should not migrate too far away—which means that it may need to be used within a certain time frame. The water must also remain of a suitable quality. Finally, there are legal limitations on who owns the recharged water. Arizona is a world leader in managed aquifer recharge, but when the Groundwater Management Act was enacted in 1980, storing water underground didn't necessarily mean one had the legal guarantee to get it back. Only through progressive laws have these issues been resolved.

California is renowned for its managed aquifer recharge. For more than seventy-five years, southern Los Angeles County has used managed aquifer recharge to overcome plunging groundwater levels and seawater intrusion from overpumping. Serving about four million people, the Water Replenishment District of Southern California recharges groundwater by using nearly a thousand acres (four hundred hectares) of spreading basins. In addition, the district recharges almost three hundred injection wells along sixteen miles (twenty-six kilometers) of coastline to combat seawater intrusion with a freshwater barrier.[3]

Some of the district's recharge water is imported from hundreds of miles away in northern California and the Colorado River. Much of it, however, comes from locally recycled water. The Water Replenishment District is the largest user of recycled water in the United States, but it's not resting on its laurels. Under the catchy slogan of WIN (water independence now), the district has ambitious plans to eliminate the need for *any* imported water for recharge by utilizing every possible drop of storm water and recycled water.[4]

Then there's Orange County. Lying between San Diego and Los Angeles, Orange County was once covered by citrus groves, but has been transformed into a sprawling suburbia with trendy seaside towns. For nearly a century, groundwater conditions in Orange County waxed and waned in response to population pressures and droughts. Today, the county is internationally renowned for its managed aquifer recharge, with the most sophisticated groundwater replenishment program in the world.[5]

The challenges of managing groundwater in Orange County began more than a hundred years ago. By the early 1930s, the area's largely agricultural economy was pumping vast amounts of groundwater from the sands and clays underlying this semi-desert coastal plain. Artesian wells, once common in places such as Fountain Valley, had gradually disappeared. In 1933, the California state legislature established the Orange County Water District to protect and manage the large underlying groundwater basin and the county's water rights to the Santa Ana River. During its first decade, the area received above-average rainfall, making the newly formed district's job easy. After World War II, however, successive years of below-average rainfall, along with growing water demands, caused groundwater levels to fall below sea level. Ocean water moved into the aquifers and coastal wells had to be abandoned.

In the 1950s, the county's population soared and Disneyland rose out of Anaheim's fields and orchards. Large increases in water demand caused groundwater levels to fall to their lowest point ever. Saltwater was found in aquifers as far as five miles (eight kilometers) inland from the ocean. Between 1956 and 1964, the situation stabilized a bit as the district's recharge program outpaced the rate of extraction. But saltwater intrusion was a continuing menace, and the costs of imported water were escalating.

In the mid-1970s, Orange County completed a twenty-first-century, state-of-the-art treatment plant for recharging treated wastewater effluent. Water Factory 21, as it was named, sounds like one of those dismal factories out of the old Soviet era, but the truth was just the opposite. Water Factory 21 was among the first facilities anywhere in the world to utilize reverse osmosis for ultra-purified water treatment. To make the point, punch made with recycled wastewater was served at the dedication ceremony.

Water Factory 21 was eventually replaced by what is today the world's largest wastewater purification system for indirect (you recharge it first) potable (you can drink it) reuse. Known by the unassuming name of Groundwater Replenishment System, it takes highly treated wastewater that previously would have been discharged into the Pacific Ocean and purifies it through a three-step advanced treatment process.

The first step involves microfiltration, where the water is drawn through polypropylene hollow fibers that are similar to straws, but have tiny holes in the sides one three-hundredth the diameter of a human hair. Through this process, suspended solids, protozoa, bacteria, and some viruses are filtered out of the water. The second step is reverse osmosis, in which water is forced under high pressure through semi-permeable membranes, thereby removing pretty much everything but the water molecules. The end result is so pure that minerals must be added back. The third and final step involves the water being dosed with hydrogen peroxide and then zapped with ultraviolet light. This treatment further disinfects the water and destroys trace organic compounds, such as pharmaceuticals, that may have passed through the reverse osmosis membranes.

As of May 2016, Orange County's Groundwater Replenishment System has treated 188 billion gallons (710 billion liters) of water for reuse. About half of the water is pumped into injection wells, where it serves as a barrier to seawater intrusion. The rest goes to recharge basins where the water filters through the sand and gravel, replenishing the aquifers. Among the many benefits of the system is a substantial energy savings. This process uses less than half the energy it takes to transport water from northern California, and less than a third of the energy required for desalination of seawater.[6]

Managed aquifer recharge is not a panacea for water shortages, nor is it universally applicable.[7] For a spreading basin to work, the subsurface must be permeable all the way to the water table so the water can move downward unhindered. Silt or clay layers can act as a barrier, preventing the water from reaching (recharging) the water table. Spreading basins and injection wells invariably clog during their operational life, and require specialized knowledge to maintain long-term operations. In some geologic settings, chemical interactions that occur when water is recharged can release arsenic or other toxic elements. In spite of these limitations, many utilities worldwide are adopting managed aquifer recharge as an important component of their water supply.

Managed aquifer recharge requires surplus water. In this respect, treated sewage effluent (recycled water) stands out as the only "water resource" whose global availability is increasing with population growth. Municipalities in fast-growing urban areas, particularly in semi-arid and arid regions, are realizing that treated sewage effluent is just too valuable a resource to waste. Using water only once is becoming a luxury many cities can no longer afford. Recycled water can be recharged to groundwater as part of a managed aquifer recharge program, or used directly as a source of water supply. If used directly, a separate delivery system commonly distributes the recycled water for landscape irrigation and other nonpotable uses.[8]

At least forty-three countries around the world have major water-recycling facilities. The United States far and away leads the world in the volume of water it recycles, yet only a small percentage of the nation's municipal wastewater is recycled. In other words, there's plenty of room for growth. On a per capita basis, Cyprus, Kuwait, Israel, Qatar, and Singapore have the most intensive water-recycling programs. Israel treats nearly all of its wastewater and sends it south to the Negev Desert for agricultural irrigation (to "make the desert bloom").[9]

The island nation of Singapore, home to some five million people, has no natural aquifers or lakes. Long dependent for its water on Malaysia (with which it has a tenuous relationship), Singapore now gets up to 30 percent of its water from recycling treated sewage effluent. The country has ambitions to increase this ratio to more than half by 2060. Marketed as "NEWater," most of the treated effluent is used for manufacturing, with some for drinking.[10]

Towns and cities using recycled water to supplement their water supply make up a remarkably diverse group. Among those on the vanguard are Grand Canyon Village in Arizona; the city of St. Petersburg, Florida; Windhoek, the capital of Namibia; and the town of Big Spring, Texas.

Grand Canyon Village, on the South Rim of the Grand Canyon, is home to the oldest dual-distribution system in the United States. Located on the high, semi-arid Colorado Plateau, the village's spectacular canyon overlook attracts millions of visitors each year. The challenge of securing a reliable water supply predates today's tourism.

The first settlers hauled their water by burros and later by rail, so it was only natural that residents eventually tried to tap any groundwater under their feet. An early attempt produced no water after drilling a thousand feet (three hundred meters), so the effort was abandoned. Despite a tantalizing view of the Colorado River below, the water is too sediment-choked to use. The Colorado River carries about a half million tons of sediment every day past any given point in the Grand Canyon. As the saying goes, the Colorado River is "too thick to drink but too thin to walk on."[11]

In 1926, Grand Canyon Village built a sewage treatment plant. With every drop of water considered a valuable resource, it incorporated a distribution system to deliver the treated sewage effluent to landscape irrigation and toilets. This use of onsite recycled water for nonpotable uses continues to this day. Solving the drinking-water problem eventually entailed securing water from Roaring Springs on the other side of the canyon, then installing pipelines and pump houses to move the water from the springs to the bottom of the Grand Canyon, across the Colorado River, and finally up to the South Rim.[12]

A larger-scale system is in place in St. Petersburg, Florida. Home to about 250,000, St. Petersburg, the "Sunshine City," holds the Guinness World Record for the most consecutive days of sunshine—768.

The city is also home to the first major treated-sewage-effluent reuse system in the United States, and one of the largest in the world.[13]

St. Petersburg sits on a peninsula on the west coast of Florida, between the Gulf of Mexico and Tampa Bay—the state's largest estuary. In the early 1970s, population growth began to outstrip the water supply from its municipal wells. At the same time, Tampa Bay's water quality and rich marine life were deteriorating. The bay harbors more than two hundred species of fish. Its mangrove-blanketed islands support the most diverse waterbird nesting colonies in North America, from the familiar white ibis and great blue heron to the regal reddish egret, the rarest heron in the United States. Wastewater and other pollutant discharges from St. Petersburg were threatening to destroy this ecological treasure. In an effort to simultaneously save the bay and the city's water supply, St. Petersburg upgraded its wastewater treatment plants and began an extensive water reuse program.[14]

Recycled water was initially provided only to large users, such as golf courses, parks, schools, and large commercial areas. Subsequently, the dual-distribution system was expanded to include residential irrigation in neighborhoods where a sufficient number of homeowners were willing to pay for the initial cost of a hookup. By 2009, recycled water was meeting about 40 percent of the city's total water demand.

Dual systems that separate potable from nonpotable uses, such as those in Grand Canyon Village and St. Petersburg, are basically noncontroversial. The same cannot be said for using recycled water as part of the drinking-water supply—what advocates call potable reuse and opponents call "toilet-to-tap."

Unless you live near the headwaters of a river, or draw from a clean groundwater source, much of the water we drink comes from treated sewage effluent that has been discharged into rivers, withdrawn downstream, treated to drinking water quality, and delivered to our homes. During low-flow conditions, particularly during a protracted drought, a large part of a city's drinking water can originate as wastewater from upstream communities. Rivers provide a natural purification process as a first step prior to water treatment and use. Recharging treated wastewater into groundwater provides additional cleaning as the water filters through the underlying soil and rocks. A small (but growing) number of water-stressed communities are bypassing these natural purification steps by directly using treated sewage effluent as part of their potable water supply. Windhoek, Namibia, is among them.

Lying north of South Africa, Namibia is the driest country in sub-Saharan Africa. The only rivers that flow year-round form the country's north and south borders. The Namib Desert along the western coast is one of the most inhospitable places on Earth, with scorching sands, wind speeds often approaching gale force, and an average yearly rainfall of just a half-inch (thirteen millimeters). Occasional morning fog drifts over the Namib Desert, but is quickly evaporated by the wind and sun.

One of the Earth's most astonishing adaptations is found here, in the form of the Namib Desert beetle. When the occasional fog circulates over the beetle's back, tiny water droplets accumulate on top of bumps on its armor-like shell. These water-attracting bumps are surrounded by waxy water-repelling channels. When the droplets become large and heavy enough, they roll down these channels to a spot on the beetle's back that leads directly to its mouth. Scientists are trying to mimic this feature to create air-moisture harvesting materials for use in places without access to safe drinking water—including a project to develop a self-filling water bottle.[15]

Windhoek, the capital and largest city of Namibia, is situated on a mile high (1,600-meter) central plateau between the hyper-arid Namib Desert and the Kalahari Desert to the east. This is an attractive and rapidly growing city of more than 300,000 inhabitants. Windhoek's numerous legacies from the days of German colonialism include Victorian buildings, an old fortress, and the gothic-style Christ Church. Tourists enjoy good German beer at Joe's Beerhouse before heading out into Namibia's spectacular national parks and nature reserves.

Like Namibia's desert beetle, Windhoek has developed a special adaptation for obtaining its drinking water. In 1968, it became the first city in the world to return treated sewage effluent directly to its potable water system. The effort began modestly, averaging around only 4 percent of the city's water for many years. After several treatment plant upgrades, recycled water grew to 35 percent of the drinking-water supply during normal periods, and as much as 50 percent during droughts. Research has shown no harmful health effects, and initial public opposition faded over time.[16]

A few other cities have followed in Windhoek's footsteps. In the United States, the small West Texas town of Big Spring seems like an unlikely place to be using treated sewage effluent in its drinking-water supply. But it doesn't have a lot of choice. The town's "big spring," which was originally fed by a small but prolific aquifer, dried up about ninety years ago upon the arrival of a railroad and the West Texas oil boom. The town has limited access to surface water, and in 2014, the closest reservoir was only about 1 percent full.[17]

Big Spring's sewage effluent is first treated at its old sewage treatment plant. Next, a special facility that uses reverse osmosis and two stages of disinfection treats the water again. Any water failing to meet several safety tests along the way is returned to the town's sewage treatment plant and begins the process again. Before being piped to homes and businesses, the water is blended with reservoir water. This blend is then distributed to five drinking-water facilities that serve about 250,000 people, where it is treated yet again by conventional technologies.[18]

Wichita Falls, Texas, a city of over 100,000 lying near the Oklahoma border, followed in Big Spring's footsteps to avoid running out of water during the worst drought on record. In 2014, Wichita Falls received state approval to begin blending a half-and-half mix of treated effluent and lake water at its drinking-water treatment plant. When the drought broke a year later, the project was replaced with indirect potable reuse, with the treated effluent being pumped to a lake.[19]

Aside from the formidable "yuck factor" in using highly treated wastewater for drinking water, today's advanced treatment technologies have one distinct advantage: they more effectively remove the plethora of modern chemicals—pharmaceuticals, birth control hormones, and personal care products—that we flush down our drains and toilets every day. These "contaminants of emerging concern" went largely unrecognized until the development of laboratory techniques that could detect them at miniscule concentrations (parts per billion, or less).

In 2002, the U.S. Geological Survey released the results of a study that tested 139 streams in thirty states for pharmaceuticals, hormones, and other wastewater contaminants. Contaminants of emerging concern were detected in 80 percent of the samples. Though the amounts were small, almost all of the ninety-five contaminants were detected at least once. The most frequently detected compounds included steroids, caffeine, nicotine metabolites, nonprescription pain relievers, and DEET, the active ingredient in many insect repellents. Antibiotics were detected in more than half the samples. Although the samples were collected downstream of wastewater discharges and other locations where the contaminants would most likely occur, the widespread detection of these compounds in streams across the nation was cause for concern.[20]

Groundwater is less affected by wastewater discharges than surface water, but is not immune to such contamination. Aquifer recharge using recycled water from conventional treatment, leaky sewers, and septic tanks are all possible sources of groundwater contamination.

At the time of the USGS study of streams, scientists in Berlin, Germany, were tracking pharmaceuticals in groundwater. Like many industrialized countries, pharmaceuticals are prescribed in large quantities in Germany. Some of these drugs pass through humans unchanged, or only slightly transformed, as they enter wastewater treatment plants. Conventional treatment processes remove some, but not all, of these pharmaceuticals.[21]

Although Berlin is known as a city of rivers and lakes, it uses groundwater for its public water supply. During summer periods of low water flow, more than half of Berlin's surface water may consist of treated sewage—a less-than-desirable source of drinking water. The city's wells are used to provide a natural pretreatment of the water. In a process known as bank filtration, the wells are located next to streams and lakes, and draw much of their water from surface-water sources.

As the water passes through the aquifer to the wells, natural processes remove microbes and chemical contaminants.[22]

For over a century, Berliners have used bank filtration as an inexpensive method of treating drinking water for pathogens and common chemical pollutants. The first indications of how this natural filtration system performs with respect to pharmaceuticals and other contaminants of emerging concern came by accident.[23]

In 1990, after Germany's reunification, the senate of Berlin commissioned monitoring of water collected from wells near the former Berlin Wall to test for the presence of herbicides that had been used to keep the area clear of vegetation. Among the chemicals detected was clofibric acid, a metabolite of cholesterol-lowering drugs that has a chemical structure similar to some of the herbicides tested. Follow-up studies found clofibric acid in Berlin tap water.[24]

German investigators began to examine other pharmaceuticals during bank filtration. Some degraded or were sorbed (attached) onto aquifer surfaces and so were never detected in the drinking-water wells. They found that a few, such as carbamazepine (an anticonvulsant and mood-stabilizing drug) and primidone (an anticonvulsant), passed easily through the system. In fact, carbamazepine is so widespread and persistent that it has been used in studies as an indicator of contamination by wastewater.

The bottom line is that today's profligate and careless use of pharmaceuticals, antibacterial soaps, cosmetics, fragrances, and other personal-care products is contributing to the widespread occurrence of numerous compounds (still at very low levels) in surface water, soils, and groundwater. Even traces of illicit drugs, such as cocaine, are appearing.[25] While some emerging contaminants have been linked to mutations of fish and other wildlife, the long-term effects on humans have yet to be determined. However, two things are clear. Use of recycled water is likely to grow as surface-water and groundwater resources are increasingly strained by population growth and climate change. And in today's world where human-made chemicals are used with casual, even reckless abandon, preserving the quality of groundwater is as critical as assuring its availability.

11

Poisoning the Well

An ounce of prevention is worth a pound of cure.
—Benjamin Franklin

In the late 1970s, the discovery of hazardous chemicals buried beneath houses in the Love Canal neighborhood of Niagara Falls, New York, made groundwater contamination front-page news across the United States. It wasn't long before hundreds of groundwater "plumes" were being discovered throughout North America and Europe. The more people looked, the more contamination was found. Worse yet, toxic chemicals that had been indiscriminately disposed of over decades were turning up in drinking-water wells. The public's initial indignation at Love Canal's problem turned to fear at the prospect of their own drinking water being laced with odorless and tasteless toxic chemicals.[1]

With lawsuits flying, an entire industry was soon created to locate, characterize, and attempt to clean up contaminated groundwater. Regulatory agencies, consulting firms, and industry began to hire additional staff at a breakneck pace. With a shortage of qualified experts, out-of-work petroleum and exploration geologists retooled themselves as groundwater experts.

In most cases, the contamination was the result of ignorance rather than malicious intent. People simply assumed that the chemical wastes would stay put or be cleansed through the natural processes of filtration, dilution, and biological degradation. Or they just weren't thinking at all. Ironically, early environmental legislation indirectly contributed to the problem. The focus on clean air and clean water meant that land disposal was the only remaining option.

In 1976, the U.S. Congress passed the Resource Conservation and Recovery Act (RCRA) in an effort to provide "cradle-to-grave" regulation of hazardous wastes. But what was already down under wasn't going to just get up and go away. There was also the complicated problem of trying to figure out who was responsible for cleaning it up. Multiple sources of contamination often make it extremely difficult to determine who "owns" which plume. And finally, many of the companies that were responsible were long gone, or went bankrupt trying to clean up their messes.

In 1980, the U.S. Congress passed the Comprehensive Environmental Response, Compensation, and Liability Act, commonly known as Superfund. Over the next decade, more than a billion dollars was spent on site investigations and cleanup activities.[2] Most of the money came from taxes on the chemical and petroleum industries, under the principle that the polluter pays. Plenty of work remains to be done. Smithville, Ontario, and Hinkley, California, illustrate some of the continuing challenges.

The small town of Smithville, Ontario, thirty miles (fifty kilometers) west of Niagara Falls, sits on a peninsula separating two Great Lakes—Lake Erie to the south and Lake Ontario to the north. It is an area of natural beauty known for its granaries and fine wines. The nearby Niagara Escarpment, a remnant of the outer rim of an ancient shallow sea, has been designated a World Biosphere Reserve by the United Nations.

Beginning in the early 1900s, the province of Ontario underwent intense industrial development when electro-chemical industries moved into the area, drawn by the availability of relatively cheap and plentiful energy. One result was a legacy of contaminated sites, among them the nearby Love Canal.

With hazardous wastes piling up, companies needed a place to dispose of them. In 1978, Chemical Waste Management Limited opened a facility in Smithville to collect and package liquid hazardous waste.

The company planned to ship the waste to a disposal facility in the United States. After the United States closed its border to hazardous waste shipments in 1980, however, the waste accumulated until the facility became filled to capacity in 1983.[3]

More than half of the waste was polychlorinated biphenyls (PCBs) manufactured in the United States. PCBs were widely used as coolants and lubricants in transformers and other electrical equipment. They were banned for most uses in Canada and the United States after evidence mounted that they concentrate in the food chain and are hazardous to wildlife and humans. PCBs are also extremely slow to break down in the environment.

In 1985, a group of local environmentalists jumped the fence at the Smithville facility and discovered PCB oil in storm water that had ponded at the site. In the midst of intense media and public attention, the Ontario Ministry of the Environment took control of the facility.

The contaminated storm water was just the first shoe to drop as PCB oil was discovered in groundwater beneath the site. Movement of the contaminants through fractures in the clay soils and underlying carbonate rocks had been greatly underestimated, and the contaminated groundwater was now headed directly for the municipal well supplying water to the town. The well was shut down and a six-mile (ten-kilometer) water pipeline was installed to provide an alternative source of drinking water.

In 1989, a system to pump and treat the groundwater began operating at the facility to prevent off-site migration of the contaminant plume. Various additional methods have since been tried to clean up the waste. None have succeeded. By 2014, cleanup at this one site alone had cost more than $42 million, with no end in sight.

Thanks to some trespassing environmentalists, the Smithville contamination was discovered before it intercepted the drinking-water supply. Other towns have not been so lucky.

Hinkley, California, is a small town on the Mojave Desert located two hours northeast of Los Angeles. Julia Roberts brought national attention to Hinkley's contaminated water supply in her Oscar-winning portrayal of activist Erin Brockovich and her David-versus-Goliath victory over Pacific Gas and Electric Company (PG&E).

The utility PG&E operates a compressor station in Hinkley for natural gas transmission pipelines. Natural gas must be recompressed every several hundred miles, and the station uses large cooling towers to cool the compressors. Between 1952 and 1964, hexavalent chromium ("chromium-6") was added to the water used in these cooling towers to prevent corrosion. The water was stored in unlined ponds where the chromium eventually seeped into the groundwater, leading to possibly the world's largest chromium plume. State regulators learned of the groundwater problem in 1987.[4]

After a protracted legal battle, PG&E awarded $333 million to just over six hundred residents. In 2013, seventeen years after the settlement, the chromium-6 plume appeared to be still growing. After two decades of recurring problems, people were pulling up stakes and moving away. The town's only school had closed. The only remaining business, a small market and gas station, was barely hanging on (and eventually closed in 2015).[5]

Chromium-6 can result from human activities, but it also occurs naturally from the breakdown of chromium-containing rocks. To determine the extent of PG&E's continuing culpability, it will be essential to estimate how much of the chromium-6 is naturally occurring, and determine whether the plume is growing or shrinking. An earlier study to determine the natural background level was discredited by a panel of experts.

As the town's ongoing disaster played out, Hinkley residents and PG&E agreed on one thing—they needed a trustworthy expert to get to the bottom of this question. In 2015, John Izbicki, a widely respected expert with the U.S. Geological Survey on the groundwater and geochemistry of the Mojave Desert, was hired to undertake this hydrologic sleuthing. Such studies, however, take time and the results will not be available until 2019.

Chromium contamination and the drinking-water standards for chromium have become major issues in California, complicated by controversy over how much of this toxin the body can safely handle.

Chromium-6 has long been recognized as harmful when inhaled, and air-pollution standards have been in place for many years. The safe level in drinking water, however, is far from established. In 2008, laboratory tests by the U.S. Department of Health and Human Services concluded that there was "clear evidence" that chromium-6 had caused cancer in rats and mice that drank water laced with it. In contrast, a California Department of Public Health epidemiology study of Hinkley residents from 1988 to 2008 failed to find anomalous new cancer cases.[6]

The U.S. Environmental Protection Agency (EPA) drinking-water standard for total chromium is 100 parts per billion (ppb). Total chromium includes both chromium-6 (the oxidized form) and chromium-3 (the reduced form), which have very different bodily reactions. Trace amounts of chromium-3 are actually good for you. It is an essential human dietary element found in many vegetables, fruits, meats, and grains.

The EPA uses a total chromium standard because chromium can change back and forth (through oxidation-reduction reactions) between chromium-3 and chromium-6 in water-distribution systems and in our bodies. This helps explain the skepticism about extrapolating studies of chromium-6 imbibed by mice and rats to humans. The human body converts the "bad" chromium-6 to "good" chromium-3 in our intestines, but rats and mice do not have the same mechanism. The question of how much chromium-6 in drinking water is harmful to humans hinges on the capacity of our gastrointestinal tracts to convert chromium from its toxic to nontoxic form.

While the EPA continued to study the problem, California proposed the nation's first drinking-water standard specifically for chromium-6. The proposed standard of 10 ppb is ten times stricter than the EPA's total chromium standard. (One ppb is roughly equivalent to a teaspoon of sugar in an Olympic-sized swimming pool.)

Approximately 4 percent of the public supply wells tested in California exceed the proposed standard for chromium-6, with about 10 percent of wells in the desert and the west side of the Central Valley failing to meet the standard. In many cases these wells belong to small water utilities, which means their treatment costs to meet the proposed standard will be large compared to their customer base.[7]

Critics of the California chromium-6 standard contend that it is unsupported by evidence of toxic effects on humans. At the other end of the spectrum, some environmental groups insist that the proposed California standard is much too lax. The debate is likely to continue for some time.

While there are many different contaminants of groundwater, in the 1980s a class of chemicals known as volatile organic compounds, or VOCs, took center stage. These chemically simple organic compounds are the components of gasoline, fuel oils, and lubricants, as well as the basis for many solvents, degreasers, refrigerants, and fumigants. Virtually everyone in today's society has products containing VOCs lying around their home, including paints, paint thinners, spot removers, moth balls, ant and roach killers, and so on. Large quantities are manufactured and released into the environment. VOCs are not only toxic, but once in groundwater they have a tendency to persist for a long time without breaking down. Another characteristic makes many VOCs uniquely problematic—if spilled in sufficient quantities, a portion can remain separate from water as a non-aqueous phase liquid, commonly referred to as a NAPL (pronounced "nap-L").[8]

Some NAPLs, like gasoline and fuel oil components, are lighter than water. These LNAPLs (pronounced "L-napLs") float on top of the water table. Fuel oil spills and leaking underground storage tanks (often abbreviated as LUSTs) make LNAPLs one of the most common types of groundwater contamination. More than 1.7 million underground storage tanks have been closed since the EPA began to address them in 1984. Many of these sites still have residual contamination.[9]

LNAPLs and their dissolved plumes can work their way to water-supply wells and surface-water bodies. Some, such as benzene, are known to be carcinogenic at very low concentrations. But there is some good news. Over time, most of these lighter organic compounds are broken down by microbes in the groundwater. Although LNAPLs are a serious and widespread environmental problem, they have proven less troublesome than their denser-than-water cousin—DNAPLs.

Beginning in the 1970s, Friedrich Schwille at the Federal Institute of Hydrology in Koblenz, West Germany, carried out detailed laboratory experiments of DNAPL behavior in the subsurface and was the first to present the concept of heavier-than-water solvents sinking below the water table. Unfortunately, these studies went largely unrecognized for many years.[10]

As DNAPLs sink beneath the water table, they leave residual contamination along the way. These are sometimes described as disconnected blobs and ganglia, to give an idea of their complexity. Further complicating matters, if the DNAPLs have enough mass, they can eventually pool when they reach a low permeability layer in the subsurface. DNAPLs can be extraordinarily difficult to locate and so provide a very long-term source of slowly dissolving contaminants to groundwater. Even relatively small quantities of these chemicals can result in groundwater contaminant plumes several miles long.

Among the most common and troublesome DNAPLs are chlorinated solvents used to dissolve greases and other substances. First produced in Germany in the nineteenth century, chlorinated solvents began to be used in a widespread way during World War II. By 1943, as many as thirty thousand degreasers used vapors from boiling solvents to dissolve the oils and greases from metal parts. Chlorinated solvents became essential to the manufacture and maintenance of aircraft, computers, machinery, motor vehicles, and components such as printed circuit boards. Many prevalent chlorinated solvents in groundwater are commonly known by their abbreviations, such as TCE for trichloroethylene and PCE for perchloroethylene.[11]

The high volatilities of chlorinated solvents (they easily vaporize) led to a false sense of security about their disposal. For many years, people thought they would readily vaporize into the atmosphere if they just poured them on the ground—a practice that was even recommended by various manufacturers.[12] What didn't vaporize was thought to be rendered harmless by natural processes before any residual amounts could reach the groundwater.

The bubble burst about the same time that the Love Canal crisis was unfolding. The popular use of septic tank cleaners on Long Island, New York, led to one of the first known cases of widespread TCE contamination. In 1977 alone, Nassau County residents bought 67,500 gallons (250,000 liters) of septic tank cleaners that were comprised mostly of organic solvents. Once poured into septic tanks, the solvents were on their way toward contaminating Long Island's groundwater, its only source of drinking water. In 1978, the New York State Department of Health closed thirty-six municipal wells, affecting two million people.[13]

Then out of left field came another major problem—dry cleaners. The origins of dry cleaning are shrouded in mystery. Explanations range from someone who accidentally spilled kerosene on a stained tablecloth, to a French sailor who fell into a vat of turpentine and came out with clean "duds." The earliest known dry cleaner opened for business in Paris in 1840. Dry cleaning made its way to the United States near the turn of the twentieth century.[14]

Though far from ideal, gasoline and other petroleum distillates were the first dry cleaning agents. The process was slow and costly, and clothes were left with a disagreeable odor. More problematic, the petroleum distillates were highly flammable. As one industry spokesperson pointed out, dry cleaning shops "tended to blow up a lot."[15] Many cities adopted ordinances requiring dry cleaners to locate away from highly populated areas. Insurance companies refused to cover them.

In the 1930s, carbon tetrachloride and TCE began to replace the petroleum distillates as dry cleaning agents. After World War II, PCE became the dry cleaner solution of choice for decades. These chlorinated solvents offered major advantages over the petroleum distillates. They were cheaper, worked faster, and virtually eliminated the problem of disagreeable odors. Moreover, they were unlikely to explode. By using nonflammable solvents, dry cleaners were allowed to locate in urban areas and shopping centers—providing a huge boost to the industry.

While pretty much everyone nowadays takes dropping off their clothing at the dry cleaner for granted, dry cleaners were once one of the most ubiquitous sources of DNAPL groundwater contamination.

Early shop owners paid little attention to the proper handling and disposal of the dry cleaning agents. Equipment and machinery were prone to leaks. Water containing PCE was frequently poured down the drain or dumped on the ground behind the shop. Regulations and practices have improved dramatically, but the legacy of these past practices remains. As of 2010, an estimated 36,000 dry cleaner facilities operated in the United States, of which about 75 percent (27,000 dry cleaners) have soil and groundwater contamination as a legacy of previous days.[16] This doesn't include all the dry cleaners that have moved or gone out of business.

While mom-and-pop dry cleaners have caused serious groundwater contamination, the industry pales in comparison to the U.S. Department of Defense, which has the largest number of facilities undergoing groundwater remediation in the United States, and probably the world. Once again, but on a much larger scale, chlorinated solvents were widely used and often disposed of in unlined pits. More than $30 billion has been spent cleaning up (or trying to clean up) hazardous chemicals at military bases.[17] Compounding the problem is the massive scale of contamination from industries supporting the military.

Following the outbreak of World War II, southern California and southwestern cities such as Albuquerque, Phoenix, and Tucson underwent rapid growth—fueled in large part by the massive funding for military bases, aircraft plants, shipyards, and aerospace and electronics industries. This rapid industrial growth coincided with the growing use of chlorinated solvents and other synthetic organic chemicals for cleaning machinery and other uses. As a result, large-scale and highly problematic groundwater contamination persists throughout the region to the present day. The San Fernando Valley and Tucson are prime examples.

At one time, the aquifer in California's San Fernando Valley provided drinking water to more than 800,000 residents of Los Angeles, Glendale, and other nearby cities. In the 1980s, more than half the water-supply wells were shut down because of groundwater contaminated by chlorinated solvents, primarily PCE and TCE. Chromium-6 and nitrate were later added to the list. The shutdown of these wells resulted in the loss of a substantial drinking water source in a populated area where water is already scarce. To make up the difference, the cities were forced to turn to more expensive and less reliable imported water. In 2013, after decades of limited progress in groundwater cleanup, the water utility serving the San Fernando Valley began exploring options for building possibly the world's largest treatment center to treat the groundwater for public supply.

In Tucson, a contaminant plume five miles (eight kilometers) long, primarily consisting of TCE, was discovered near the airport in 1982. This discovery followed years of complaints about health problems and a foul chemical odor in private wells from the nearby, and mostly poor, Latino community. Among the culprits were aircraft and electronics facilities that had dumped their TCE-laden wastes into unlined pits in the desert.[18]

Wells supplying water for more than 47,000 people were shut down. An expensive treatment system was installed whereby groundwater is first air stripped to remove the VOCs from the water, followed by air emission controls to keep the VOCs from venting to the atmosphere. Since 1994, more than four thousand pounds (1,800 kilograms) of VOCs have been removed. Some of the treated water is reinjected into groundwater, but much of it is added to the city water system.

In 2002, twenty years after the contamination was discovered, 1,4-dioxane was detected in the Tucson airport plume. Widely used as a solvent stabilizer, 1,4-dioxane had escaped notice by nearly everyone everywhere until the late 1990s. The chemical is a potential human carcinogen and is among the most mobile contaminants in groundwater. Removal of the 1,4-dioxane required construction of an additional state-of-the-art treatment facility.[19]

For many years, the prevailing belief was that the risk of exposure to groundwater contamination is minimal, providing people don't drink the untreated water. In the 1990s, however, an insidious new contaminant pathway was recognized. Vapors from VOC groundwater plumes were discovered to be working their way up through soils and entering homes and businesses through cracks in the foundation, and even the tiny openings for utility lines.

One of the first places to bring the vapor intrusion danger to national attention was the Redfield site in Denver, Colorado, where scopes for rifles and binoculars were manufactured from 1967 to 1998. About half of the eight hundred homes and apartments tested required some level of mitigation.[20]

The vapor intrusion pathway raised a whole new quandary. Bottled water can be substituted when drinking water has been contaminated, but indoor air pollution from hazardous vapors is much harder to deal with. Any number of scenarios is possible. For example, a children's dance studio could be in the same building as a long-gone and forgotten dry cleaner, with teachers and children being exposed to dangerous levels of airborne VOCs from an underground PCE plume. The plume might even have been believed to be satisfactorily contained. Closed cases suddenly needed to be reexamined.

Vapor intrusion is yet another of the many ways in which some portion of a contaminant plume migrates along a preferential pathway into the biosphere. These pathways are nearly always impossible to predict.

In 1999, while commenting on the problems with the gasoline additive methyl tertiary butyl ether (MTBE), EPA administrator Carol Browner declared that Americans should have "both cleaner air and cleaner water—and never one at the expense of the other."[21] MTBE in gasoline challenged this ideal. The story of its use begins with a problem familiar to many older car owners.

Engine knock occurs when the gasoline-air mixture in a motor's combustion chamber explodes prematurely before the spark plug fires. To eliminate engine knock, the octane rating of gasoline must be high enough. This was achieved by adding tetraethyl lead to gasoline, until concerns about lead in the environment resulted in its phase out in the 1970s. (Lead also interferes with catalytic converters in modern cars.)

Despite its name, methyl tertiary butyl ether is a simple organic compound made up of carbon, hydrogen, and oxygen, and appeared to be the perfect substitute for lead in gasoline. MTBE has a high octane rating, low production cost, and blends easily with gasoline. This seemingly made-to-order additive was introduced into gasoline in 1979, but its use took off with passage of the Clean Air Act amendments of 1990. These amendments required that areas with severe ozone-smog problems add an oxygenate like MTBE to gasoline to promote more thorough burning in engines. Other areas soon adopted this concept. By 1999, MTBE was the second-most-produced organic chemical in the United States.[22]

While trying to solve an air pollution problem, it wasn't long before a new problem emerged—groundwater contamination. In retrospect, it's not difficult to understand why. MTBE dissolves easily in water and tends not to attach (sorb) onto soil and aquifer material, which means that it can travel quickly through soil and groundwater. MTBE is also slow to biodegrade; although it is now known to be less persistent than originally thought.[23] Potential sources of MTBE to groundwater are everywhere, most notably leaking underground storage tanks, but also leaks during fuel transport, gasoline spills by homeowners, storm water runoff, and even washout from the atmosphere by precipitation.

MTBE is a possible carcinogen, although the human health effects have not been clearly established. What really brought MTBE to the forefront was that, even at very low concentrations, it makes water smell and taste foul and nauseating. The closest comparisons are rubbing alcohol and turpentine.

In 1997, the EPA issued an advisory that MTBE concentrations in drinking water should be kept below 20 to 40 ppb.[24] This miniscule amount would generally avoid unpleasant taste and odor effects, although some people are sensitive to MTBE at concentrations as low as 1 ppb. The threshold of 20 ppb—equivalent to one drop in five hundred gallons of water—was considered to be protective of human health while further studies of the risks were being conducted.

The first major hit to a public water supply system came in 1996. The city of Santa Monica, California—famed for its beaches and wealthy enclaves of Hollywood icons—discovered that it was pumping groundwater with MTBE concentrations as high as 600 ppb. Two well fields, amounting to half the city's drinking-water supply, were shut down. The city was forced to purchase replacement water. A year later, contaminated municipal wells at South Lake Tahoe, California—bordering one of the most pristine and gorgeous lakes in the world—brought further national attention to the problem.

Despite its relatively short history of use, MTBE became the second most frequently detected organic compound in drinking water wells. The chemical was particularly prevalent in drinking water in the highly populated New England and Mid-Atlantic States, where its use was correspondingly high. Most of the concentrations were low, but the relatively high rate of detection showed how vulnerable many aquifers are. A key lesson was how rapidly a commonly used chemical could cause widespread groundwater contamination.[25]

In 1999, an EPA-appointed Blue Ribbon Panel on oxygenates in gasoline recommended that the oxygenate requirement for gasoline in large cities be dropped and MTBE use in gasoline be substantially reduced. This same year, California ordered oil companies to phase out MTBE in gasoline by 2002. By 2004, nineteen states had enacted legislation to completely, or partially, ban MTBE use in gasoline. The Energy Policy Act of 2005 dealt the final death blow to MTBE use in the United States by completely eliminating the oxygenate requirement. MTBE continues to be used in Asia and elsewhere around the world.[26]

Most groundwater remediation (cleanup) projects from the late 1970s to the 1990s relied on pump-and-treat systems. The idea was straightforward—pump the contaminated groundwater out, treat it, and then inject it back into the aquifer. Pump-and-treat systems were intended to halt the spread of contamination and clean up the plume.

It was initially believed that the main limitation to pump-and-treat was the tendency of chemicals to sorb onto the surfaces of the rocks through which the water flows. Designers of early pump-and-treat systems assumed that by flushing a large amount of clean water through the aquifer, almost all of the dissolved and sorbed contaminant would be removed. They were mistaken. After an extended period of pump-and-treat, it wasn't unusual to discover that the concentrations of contaminants quickly rebounded.[27]

Groundwater scientists began to appreciate the importance of diffusion as a major problem with pump-and-treat systems. In the same way that a few drops of food coloring spread out in a glass of water, contaminants diffuse from higher to lower concentration areas in groundwater. Large amounts of contaminants can leave the main path of groundwater flow and become trapped in low-permeability and stagnant zones. While diffusion attenuates the concentrations in the plume, it also significantly extends the time required to clean up groundwater.

Contamination of fractured rocks, such as the carbonate rock at Smithville, presents even greater challenges. The interconnected network of fractures provides the main pathway for groundwater flow, but diffusion can transfer a significant mass of the contaminant from the fractures into the rock matrix itself—which then becomes a long-term source of groundwater contamination.[28] Contaminated fractured bedrock aquifers are widespread throughout the world.

British hydrogeologist Stephen Foster first recognized the importance of matrix diffusion in the early 1970s during investigations of the Chalk Aquifer, an important source of water supply in Britain. Thermonuclear testing had increased tritium levels in precipitation, and the bomb-level tritium was expected to move rapidly through fractures in the rock. When relatively low levels of tritium in the Chalk Aquifer contradicted these expectations, Foster presented evidence that diffusion from the fractures to the rock matrix could explain the delayed pulse of tritium.[29]

The ineffectiveness of pump-and-treat for cleaning up contaminated groundwater has led to many innovative alternative or supplemental approaches. Multiple techniques may be applied over time as the contaminant cleanup evolves. Initially, considerable attention is placed on treating the source zone of the contaminants. A key challenge, however, is that the source zone is almost always highly uncertain—particularly for DNAPLs, with their multiple "bull's-eyes" of contamination.

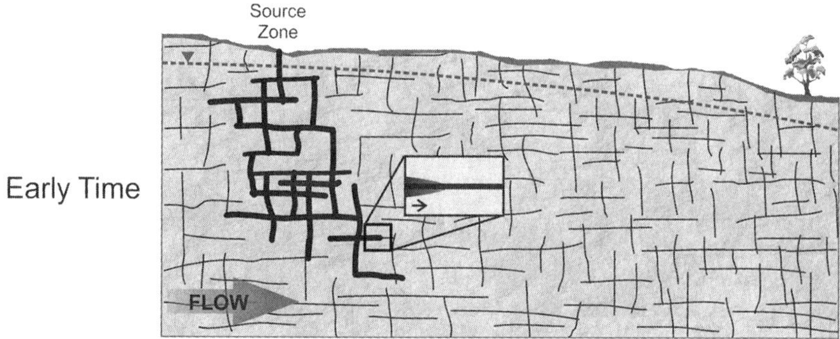

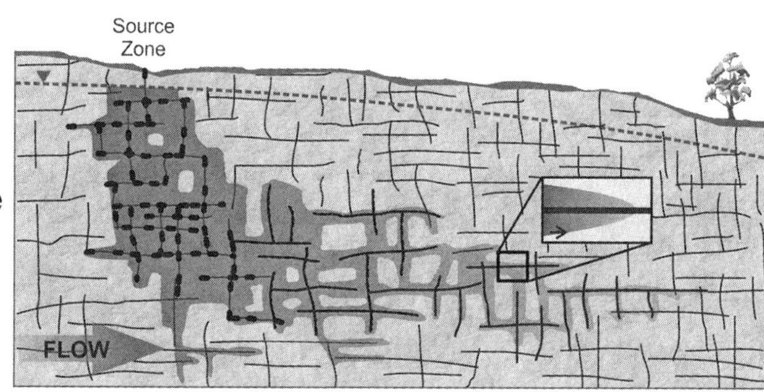

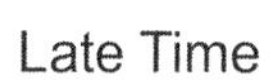

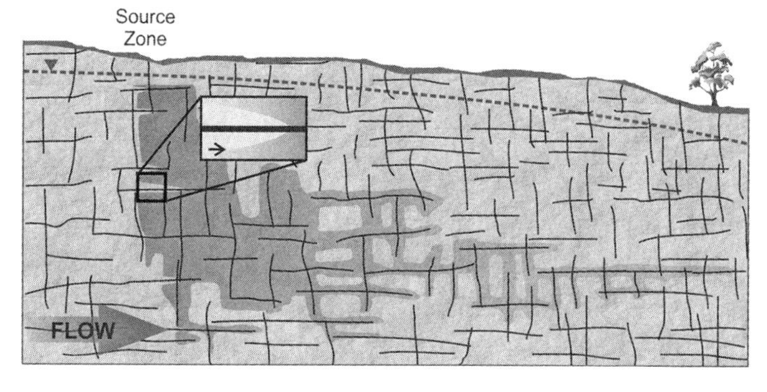

Conceptualized contamination of fractured sedimentary rock by DNAPL. *Early time:* DNAPL flows into fracture network. *Intermediate time:* Dissolved contaminant diffuses into the rock matrix along fractures containing DNAPL and where the dissolved phase has been transported downgradient in fractures. *Late time:* DNAPL has completely dissolved and nearly all the mass resides in the rock matrix, serving as a long-term source for persistent groundwater contamination. Provided by and with permission from Dr. Beth L. Parker, Professor in the School of Engineering, University of Guelph, Ontario, Canada.

Soil vapor extraction is a common method for treating soils contaminated with VOCs. In this approach, a vacuum is applied to a system of wells to pull the contaminants to the surface as a vapor (gas), where they are captured and treated. A similar approach for capturing VOCs in shallow groundwater, known as air sparging, injects air into the saturated zone.

A downside to pump-and-treat and other extraction techniques is that they can generate large volumes of waste material at the surface. In contrast, in-situ approaches work within the aquifer, thereby minimizing waste. In a common approach, oxidizing agents such as hydrogen peroxide are injected in order to transform contaminants into nontoxic chemicals such as carbon dioxide and chloride.

Among the more effective remediation approaches are those that heat the aquifer. These thermal technologies range from those that vaporize contaminants, to high-temperature techniques that trap them in vitrified (solidified) form.

In another in-situ method, hydrogeologists install a permeable reactive barrier (by digging trenches or injecting reactive material into the subsurface through boreholes) to treat groundwater as it flows through the barrier. The most widely used permeable reactive barriers are made of granular iron, which degrades chlorinated solvents such as PCE and TCE.

Biological approaches are particularly popular. These approaches enlist nature's help in the form of microorganisms that break down or degrade hazardous substances into less toxic, or innocuous, substances. Like humans, microorganisms eat and digest organic substances for nutrition and energy. This natural process can be enhanced by adding microbes capable of degrading specific contaminants, or by adding substances to stimulate biodegradation. Large amounts of everything from emulsified vegetable oil to high-fructose corn syrup have been injected into contaminated aquifers for remediation.

Under the right conditions, the natural system can cleanse itself. While this approach obviously saves money, it needs considerable evidence that it actually works to gain public acceptance. Monitored natural attenuation, as this approach is known, requires periodic checks to see how the system is performing.

Over the past forty years, billions of dollars have been spent trying to clean up contaminated sites. Despite this investment, at least 126,000 contaminated sites in the United States require continued remediation efforts for the indefinite future. Restoration of many complex contaminated sites is not possible in less than fifty to a hundred years—if even then. The eventual price tag for this undertaking is estimated to exceed $110 billion. No one really knows the true costs, not to mention all the yet-to-be-discovered sites.[30]

There is also sharp disagreement over the question of how clean these sites should be. For many years the goal has been to clean groundwater to drinking-water standards. Yet over time, it has become increasingly clear that remediating an entire groundwater plume to drinking-water standards is often a daunting, or just plain impossible, task. Given these problems, when does one declare the cleanup a success?[31]

Similar, if not worse, contamination occurs throughout the world. Most of India's approximately 2.6 million small-scale industries do not treat their industrial discharge. High concentrations of heavy metals such as chromium, nickel, cadmium, and mercury have been found in groundwater throughout the country. The situation in China is no better. In 2014, the state media acknowledged that nearly 60 percent of the country's groundwater is polluted.[32]

Virtually all chemicals that humans release into the biosphere can find their way into groundwater. The cost of protecting an aquifer through proper handling and disposal of hazardous chemicals is miniscule compared to the cost of trying to fix a contaminated aquifer—if it can even be done. As noted by John Cherry, Canada's preeminent contaminant hydrogeologist, once groundwater is contaminated you often "just have a patient on life support."[33]

12

Pathogens

More people die from unsafe water than from all forms of violence, including war.
—U.N. Secretary-General Ban Ki-moon

A 2013 issue of the *Lancet,* one of the world's most prestigious medical journals, contains a remarkable admission. After what it called "an unduly prolonged period of reflection," the journal wished to revise its obituary for John Snow. More than 150 years had passed since the journal had recognized Snow's passing in a single sentence: "His researches on chloroform and other anaesthetics were appreciated by the profession."[1]

Snow is considered one of the fathers of modern epidemiology. He is perhaps best known for persuading authorities to remove the handle from a contaminated well pump during a London cholera outbreak in 1854. Snow followed this action with epidemiological evidence of the connection of cholera to contaminated drinking water. His findings inspired fundamental changes in the water and wastewater systems of London and other cities, leading to a significant improvement in general public health around the world. Eventually, Snow's linkage of cholera to water contamination was a major reason that cities throughout the United States and Europe turned to groundwater as a safer source for their water supply.

The brevity of the obituary and failure to even mention cholera reflect a feud between Snow and the *Lancet*'s founding editor, Thomas Wakley. Their debate centered on whether miasma, the stench from decaying vegetable and animal matter from trades such as tanning and soap boiling, was responsible for epidemic disease. Wakley was sure of this connection, as were most medical men at the time. After examining weekly deaths in London, Snow instead concluded that cholera was a disorder of the digestive system that spread from fecal matter through contaminated drinking water.

The vast majority of microorganisms found in water do not cause disease. Those that can are called pathogens. Groundwater contains fewer pathogens than surface water, yet the biological integrity of groundwater cannot be taken for granted.

Pathogens do not persist indefinitely in the environment. Researchers commonly measure the rate at which their populations decline in half-lives—the time taken for a 50 percent reduction in number. Half-lives of pathogens vary from a few hours to many months, depending on the pathogen and environmental conditions such as temperature, moisture content, and acidity (pH).

The reduction in pathogens occurs by several mechanisms. Pathogens can be consumed by other organisms in the subsurface, and they can be adsorbed or filtered as they travel through soils and groundwater. Adsorption occurs when the microorganisms become attached to particles, which removes them from the water or at least delays their transport. Filtration results when the pathogens are too large to fit through the aquifer pores and cracks. The extent of filtration depends on the type of soil and rocks through which groundwater flows. For example, silts are more effective at trapping microorganisms than sands, and sands are more effective than gravel. Filtration reduction also depends on the size of the organisms. Viruses readily pass through many pores and cracks. Protozoa, such as *Giardia* or *Cryptosporidium*, are larger and are typically trapped in common aquifer materials. Groundwater contamination by these larger pathogens is often limited to wells that are flooded, or otherwise contaminated, by surface-water infiltration. Bacteria fall somewhere in between viruses and protozoa in size, and can travel through groundwater under the right conditions. These various factors lead to recommendations for minimum separation distances between drinking-water wells and septic tanks or other sources of pathogens.

Groundwater contamination also can occur from poor well design and construction. A proper sanitary seal around the well casing is essential to block contaminants that might migrate from the land surface down the outside of the casing (well annulus) to the water table. As a result of poor well seals, contaminants can rapidly bypass the unsaturated zone that naturally helps cleanse groundwater.

Detecting a full suite of pathogens in water samples on a routine basis is very expensive, due to the large number of pathogens and their episodic occurrence. For this reason, microorganisms that are indicators of fecal contamination (for example, *E. coli*) or general microbial abundance (such as total coliform bacteria) are commonly checked instead. Most coliform bacteria pose no threat to humans, which makes total coliform a weak indicator of bacterial contamination. The presence of coliform bacteria, however, can indicate declining water quality. *E. coli* is the only member of the total coliform group that is found exclusively in the feces of humans and other animals that can survive in the environment. For this reason, it's a better (though far from perfect) indicator of the possible presence of other pathogens.

Although groundwater is typically a safer source of water than surface water, microbial contamination of groundwater is problematic worldwide in developing countries, and a major consideration in the siting and construction of water-supply wells. But the problem is not limited to developing countries.

Over the period 1971 to 2006, contaminated groundwater was implicated in just over half of the disease outbreaks related to drinking water in the United States. Many of these outbreaks were from wells that serve businesses or from small water systems that do not require water disinfection and have minimal microbial monitoring requirements. People drinking from wells or springs that serve single families also can become exposed to waterborne pathogens, yet outbreaks from private wells are rarely reported. The National Ground Water Association and many states recommend annual testing of private wells, analogous to an annual health checkup with a doctor.[2]

In May 2000, in Walkerton, Ontario, a virulent strain of *E. coli* killed seven people and made more than two thousand others seriously ill. All told, the bacteria affected almost half the population of this small farming community located about a hundred miles (160 kilometers) northwest of Toronto. The shock that a community water supply could cause death and illness in Canada's largest province at the beginning of the twenty-first century made national headlines in Canada and the United States. A public inquiry was commissioned to examine what happened and determine how to prevent it from occurring again.[3]

The disease outbreak had begun after a heavy spring rainfall (a one-in-sixty-year storm) caused one of the city's shallow wells to become contaminated by pathogens from manure on an adjacent farm. How the pathogens traveled from the farm to the well was never fully established, but the high pumping rates of the well and the karst (fractured limestone) conditions of the shallow aquifer allowed rapid transport once pathogens had reached the groundwater. The operator of the farm had been following best farm practices and was exonerated by the inquiry.[4]

The contaminated well had long been recognized as vulnerable to contamination. After detecting fecal coliform when the well was commissioned in 1978, the hydrogeologist's report had stressed the need for chlorination and recommended that the town purchase a buffer zone to protect the well. The town took no action on this last recommendation. The Ontario Ministry of Environment required chlorination, but practices at the facility were sloppy. The system supervisor was supposed to measure the chlorine residual daily, yet failed to do so on most days and recorded fictitious entries on the daily operating sheets.

The contamination most likely entered the well on May 12, a week before illness became evident in the community. When asked on May 19 (and again on May 20) whether there were any problems with the quality of the town's drinking water, the general manager of the system assured local health authorities that the water was satisfactory—

despite having received adverse microbiological monitoring results a couple days earlier.

On Sunday May 21, the Ontario regional medical officer and his staff began marking a town map with a yellow dot for each diarrhea case. The map was yellow by the end of the day. "We knew there was only one thing that can do that—the water supply," he reported.[5] Health officials immediately issued a boil-water advisory, but it was too late. The first victim died the next day. At least eight days without valid chlorine residual monitoring had passed between the well contamination and the boil-water advisory.

The inquiry revealed failures at many levels, including poor system management and operations, inadequate operator training, inadequate watershed protection, and ineffective regulatory oversight by the province of Ontario. Though the operators had lied and falsified records, they had no idea of the risks they were bringing upon their community. (They continued to drink the water themselves during the outbreak.) Not knowing the source of contaminants, the local hospital made the situation worse by recommending that parents encourage their children with diarrhea to drink more fluids, thereby increasing their exposure to the contaminated water.

The health crisis in Walkerton caused Ontario and other provinces to take a hard look at their drinking-water safety through better monitoring, enforcement, training, and source-water protection programs. Despite an overall abundance of surface water, nearly 30 percent of Canada's population depends on groundwater for its drinking-water supply and more than 80 percent of the country's rural population relies on groundwater for its entire water supply.[6]

Human enteric viruses (those that replicate in the intestinal tract of humans) are the microbial contaminants of greatest concern in well water. These tiny pathogens can be excreted in enormous numbers (trillions) in small amounts of feces from infected people, do not need food for survival, and are capable of surviving in the environment outside of their host. In some cases, only one to ten viruses of the many trillions excreted is needed to cause acute gastrointestinal illness—vomiting, diarrhea, and potentially death. This relation between high numbers excreted and very few needed to cause new illness is a primary reason why virus-related illness is so easily transmitted.

Methods for detecting viruses have greatly improved, and researchers have detected nearly every one of the more than one hundred known human enteric viruses in groundwater.[7] Yet the public health risk of viruses in groundwater is still not well understood. Studies of viruses in Wisconsin groundwater have provided some surprising new insights.

During a multiyear period, researchers collected virus samples from six municipal wells tapping sandstone aquifers beneath Madison, Wisconsin. The wells are deep, ranging from approximately seven hundred to a thousand feet (220 to three hundred meters) in depth. Three of these wells draw water from beneath a regional shale confining unit, which was normally assumed to protect the underlying groundwater from the downward transport of viruses from sources near land surface. Earlier studies of two such wells in southern Wisconsin had suggested otherwise, motivating the study.[8]

The researchers collected water samples from the six wells every two to four weeks for a total of twenty-six sampling events. Viruses were detected at least eight times in all six of the sampled wells. In addition, strong correlations were discovered between virus serotypes (distinct variations within a species of viruses) in the well water and in the sewage flowing to the local wastewater treatment plant. In several cases, occurrence of a "new" virus in the sewage was followed, within weeks, by detection of the same virus serotype in municipal well water.[9]

Untreated sewage leaking from sewer pipes was clearly the most likely source of the viruses in the well water. The presence (or absence) of identical serotypes in wells and sewage at roughly the same times strongly suggested very rapid transport (days to weeks) between the sewer lines and the groundwater system. This finding was remarkable, given the depths of the well casings and that three of the wells drew water solely from beneath a confining unit. The rapid transport occurred because the large pumping rates had quickly pulled contaminants through fractures and other preferential flowpaths in the subsurface. The results suggest that other urban areas with aging, leaking sanitary sewers in proximity to public supply wells could also be at increased risk of wastewater-derived contaminants entering the municipal water supply.[10]

About the same time that these groundwater studies were under way, infectious disease specialists designed an experiment to try to answer the basic question of how many fewer cases of gastrointestinal illness might occur by disinfecting the municipal water supply. The answer is relevant beyond Wisconsin. In the United States, about twenty million people drink from public groundwater systems that have no disinfection.[11]

The researchers received permission from fourteen Wisconsin communities that had no disinfection in place to install ultraviolet (UV) light disinfection on their municipal wells. Viruses rapidly degrade in the presence of UV light, thus ensuring that the municipality was producing virus-free water during this part of the study. During the first year, eight of the communities had the UV disinfection installed. The other six communities served as controls. In the second year, the UV disinfection units were switched. More than six hundred households completed a weekly symptom checklist that they mailed to the study team.[12]

All fourteen communities had human viruses in their municipal well water at times (one in four samples were virus positive), and occasionally at very high levels. There was a strong relationship between the levels of viruses measured in household tap water and rates of illness in the communities. The type of virus most strongly related to illness was norovirus—the same virus known for causing outbreaks on cruise ships. When the UV disinfection was in place, the overall reduction in acute gastrointestinal illness among the communities was 13 percent. Once again, leaking sewers appeared to be the source of the viruses, as further evidenced by finding detergents, cholesterol, and other wastewater indicators in the well water.[13]

With these findings as a backdrop, interest turned to the disinfection requirements of Wisconsin municipalities. Instead of requiring disinfection for all groundwater systems nationwide, the EPA uses a risk-based strategy known as the "groundwater rule" to target groundwater systems that are susceptible to fecal contamination.[14] The rule mandates well-water monitoring for fecal indicators and corrective action when these are detected. The results of annual monitoring at the municipal water systems in the Wisconsin studies had satisfied the EPA requirements for nondisinfection. Yet the findings of illness caused by sporadic virus contamination raised questions about whether disinfection should be required for all municipal wells in Wisconsin, including those exempted under the federal EPA regulations.

In December 2010, the Wisconsin Department of Natural Resources ruled that sixty-six municipal groundwater systems that were not disinfecting their water had to begin doing so within three years. Though over 90 percent of the Wisconsin municipal systems already disinfect their water, some of the newly regulated municipalities balked at the idea due to perceptions of high disinfection costs. A price tag of $2.9 million was cited for one community to comply with the mandate, but this was far more than the typical cost of $10,000 per well for chlorine disinfection.[15]

In 2011, one of the first acts of newly elected Governor Scott Walker and the Republican-controlled legislature was to pass what Democrats called the Dirty Drinking Water Act, which prohibited Wisconsin's Department of Natural Resources from mandating disinfection beyond the EPA requirements. Mark Borchardt, the lead researcher on the disinfection study, testified in favor of requiring disinfection to reduce waterborne disease transmission. He noted that if the annual baseline rate is 1.2 episodes per person, then a 13 percent reduction in acute gastrointestinal illness translates to sixteen thousand fewer illnesses for a population of one hundred thousand. Borchardt also argued that disinfecting made economic sense. Over a five-year period, he estimated the cost of disinfection at roughly $1.3 million for all sixty-six systems, while saved healthcare costs would be at least $2.7 million. The actual savings would be higher, as the estimate only included direct payment to healthcare providers—it did not include the costs of lost work by the ill person or their caregiver. Borchardt's argument went unheeded. Currently, Wisconsin does not require disinfection of groundwater for municipal systems.[16]

13

Arsenic

A good working definition of a public health catastrophe is a health effect so large even an epidemiological study can detect it.
—David Ozonoff

From the time of the Roman Empire through the Renaissance, arsenic was the poison of choice. It was used to settle old scores, remove a relative standing in the way of riches or advancement, execute criminals, and commit suicide. Several features contributed to arsenic's popularity as a homicidal agent. It is colorless, odorless, and tasteless when mixed in food or drink. It's easy to obtain. Symptoms of arsenic poisoning mimic food poisoning and other common disorders. Given as a series of small doses, arsenic produces a subtle form of chronic poisoning characterized by a loss of strength, confusion, and paralysis. Poisonings were said to be so common that few people believed in the natural deaths of princes, kings, or cardinals. The French referred to arsenic as the *poudre* (powder) *de succession*.[1]

In 1836, an English chemist named James Marsh perfected a sensitive chemical test for arsenic, continuing a steady improvement of detection (and decline in use) that had started in the eighteenth century.[2] As the sinister uses of arsenic decreased, however, commercial applications of the element became increasingly common. Arsenic compounds appeared in pesticides, rat poisons, wood preservatives, and as an additive to poultry and swine feed. Arsenate sprays used against pests such as the gypsy moth and the boll weevil were among the most efficacious of their time. Although localized contamination from these uses remains, today's environmental problems are primarily linked to arsenic's natural occurrence.

Arsenic is commonly concentrated in sulfide-bearing mineral deposits and several other geologic settings. It has a strong affinity for pyrite, one of the more ubiquitous minerals in the Earth's crust. Geothermal activity, volcanic rocks, and organic-rich black shales are all significant natural sources of arsenic. Depending on groundwater pH, oxidation-reduction conditions, temperature, and solution composition, arsenic can be easily mobilized in groundwater.[3]

In the 1960s, evidence emerged in Argentina that arsenic in drinking water might cause cancer, particularly in the lung and urinary tract. Startling results from Taiwan, appearing in 1985, showed increased mortality from several cancers. Today, arsenic is a known carcinogen linked to numerous forms of cancer. Exposure to low levels also can boost risks for diabetes, heart disease, and immunological problems.[4]

Arsenic contamination occurs in many areas around the globe. In addition to Argentina and Taiwan, high-arsenic groundwater areas occur in Chile, China, Hungary, Mexico, and Vietnam. By far the most infamous case, however, involves groundwater-induced arsenic poisoning of villagers in the Bengal Basin of Bangladesh and West Bengal, India. Described as "the largest poisoning of a population in history," this unintended calamity began innocently enough and with the best of intentions.[5]

In 1949, the Environmental Hygiene Committee of India's newly formed government estimated that cholera, dysentery, and diarrhea were responsible for more than four hundred thousand deaths annually in India.[6] Bangladesh was suffering a similar fate. To surmount these problems from waterborne disease, the World Health Organization (WHO) and the United Nations Children's Fund (UNICEF) proposed the large-scale use of tube wells. These wells consist of tubes, typically two inches (five centimeters) in diameter, inserted into the ground to depths of around sixty-five to 130 feet (twenty to forty meters). The tubes are equipped with a cast iron or steel hand pump.

Officials anticipated that these wells would be relatively free of the contaminants plaguing both surface water and the shallow wells dug by villagers.

Many communities were at first reluctant to use the tube wells. Some villagers called it "devil's water," believing that demons resided underground. Local officials and aid agencies, however, convinced the communities that the groundwater was safe. As many as ten million tube wells were installed to provide drinking water for more than 130 million people. Unfortunately, no one thought to test for arsenic.[7]

In 1983, K. C. Saha, a dermatologist in Calcutta, the capital of West Bengal, was apparently the first to notice arsenic-induced skin lesions on his patients. The lesions included thickening and discoloration of the skin, and growths on the hands and feet. Saha and others began to search for a cause. By 1993, it was clear that the problem was related to the high levels of arsenic in the drinking water from tube wells, and was widespread throughout Bangladesh and West Bengal. A major impediment to a quicker diagnosis is that arsenicosis, as the disease is known, often takes a decade or longer for the tell-tale signs to appear.[8]

Around this time, Dipankar Chakraborti, at the School of Environmental Studies of Jadavpur University in Kolkata (Calcutta), began sending letters to the Bangladeshi government, UNICEF, WHO, and anyone else who might listen, highlighting the arsenicosis problem. At first, the government bureaucracies and international organizations dismissed Chakraborti as an alarmist. They considered the problem relatively minor compared to the benefits of the sharp drop in mortality rates due to diarrhea and other waterborne diseases that had come from switching to the tube wells. It was not until the late 1990s that the arsenic issue was finally taken seriously. The delay between the early identification of local cases of arsenicosis and widespread testing of groundwater supplies for arsenic proved severely costly to the health and welfare of millions of people across the Bengal Basin.[9]

There were also huge social costs. People with arsenic poisoning in Bangladesh and India suffer enormous social stigma and are shunned within their villages. Many people believe that arsenicosis is contagious or a curse. Parents are reluctant to let their children play with children who are suffering from arsenic poisoning. Married women with arsenicosis face the risk of divorce or abandonment in the male-dominated society. There is no known cure for chronic arsenic poisoning, although vitamins and better nutrition (if they can be afforded) can help in the recovery from skin lesions.

A characteristic feature of areas with high arsenic in groundwater is the large degree of well-to-well variability in concentrations. Water with dangerously high arsenic concentrations can be found in wells very near those with safe water. This capriciousness left no alternative but to analyze each well. As officials tested millions of tube wells in India and Bangladesh, the results were dumbfounding. Concentrations above the WHO drinking-water guideline of 10 ppb (parts per billion) were found to threaten the health of an estimated seventy million people, 30 percent of the combined population of Bangladesh and West Bengal.[10]

Government officials painted arsenic contaminated wells red and "safe" wells green, with "safe" being defined as 50 ppb—five times greater than the WHO guideline. Despite attempts to educate communities about arsenic contamination, large numbers of people continued to drink from the same contaminated water sources. This was particularly a problem where only a few villagers displayed any signs of arsenic-caused disease and the water appeared clear and clean.[11]

In many arsenic-affected areas, few safe water options were available. It was also hard to compete with the low-cost, easy maintenance, and convenience of shallow tube wells. Filtering well water to remove arsenic is one option, yet it is not feasible over the long term in countries such as Bangladesh with a per capita income of just a few dollars a day. Filters are relatively expensive, cumbersome to maintain, and require regular testing to ensure timely replacement. An alternative solution favored by governments was treatment of surface water through community systems, but most communities preferred groundwater and were unwilling to switch back. While various groups continue to work on low-cost treatment systems to remove arsenic, the only remaining option was to "dig deep."[12]

Excessive arsenic concentrations are largely restricted to the uppermost three hundred to five hundred feet (one hundred to 150 meters) of the Bengal Basin. Groundwater modeling studies suggest that deeper aquifers could serve as an arsenic-safe domestic water supply for hundreds, or even thousands, of years. Yet even these wells must be closely monitored and designed to avoid leaks of contaminated water from shallow to deeper aquifers. There's also a catch—water from deeper wells is predicted to remain safe only if irrigation water, with its much larger water demands, is not pumped from the deeper zones. The deeper aquifers need to be reserved for domestic uses only.[13]

The sources of the region's arsenic are coal seams and sulfide minerals eroded by streams and rivers draining the Himalayas. The sediments were deposited by the Ganges, Brahmaputra, and Meghna river systems, and their precursors. When exposed to the atmosphere, the arsenic chemically transferred (oxidized) from the sulfide minerals to iron oxides. Many of the aquifer sediments are now capped by a layer of clay or silt, limiting air entry to the aquifers. The reducing (low-oxygen) conditions favor the release of arsenic that is bound to the iron oxides. Overall, the basic geochemical factors determining arsenic release are largely understood: a lack of oxygen and the presence of organic carbon (of the right type) to feed and energize those bacteria that enhance the release of arsenic from the mineral surfaces.[14]

Less certain is the extent to which human activities have contributed to the region's arsenic problem. For over four decades, large numbers of irrigation wells have enabled Bangladesh to become self-sufficient in its food production, even though the population nearly tripled. This large-scale irrigation pumping dramatically altered groundwater flow patterns in complex ways.[15]

Human-made ponds also have a role. When Bangladesh villagers build a house, they often dig a pond. Residents use the excavated clay to elevate the house above floodwaters, while they use the pond for raising fish and bathing. Prior to tube wells, the pond water was also a drinking-water supply. Studies show that groundwater recharge from the ponds carries reactive organic carbon into the shallow aquifer, thereby fueling reactions that release arsenic from the sediments.[16]

While the Bengal Basin has been the most studied and has the most chronic arsenic poisoning, other densely populated alluvial and deltaic plains in Asia have similar problems. Estimates of the rural population in Asia exposed to unsafe arsenic levels by drinking untreated groundwater have grown to over one hundred million. The Red River in northern Vietnam and the Mekong Delta of Cambodia and Vietnam both have high levels of arsenic in groundwater. Although less studied than the Mekong Delta, groundwater beneath the Irrawaddy Delta of Myanmar is suspected to have high arsenic concentrations as well.[17]

Studies of the Mekong Delta have revealed arsenic contamination in deeper groundwater, particularly for older wells. Stanford researchers hypothesize that compaction of interbedded clays from intense pumping (causing land subsidence) is expelling water containing dissolved arsenic or arsenic-mobilizing organic carbon.[18] This mechanism for arsenic release presents yet another cautionary tale on the importance of restricting deep groundwater extraction to low-volume uses in Asia's deltas.

Skin lesions and other obvious signs of arsenic poisoning have been observed much less in Southeast Asia than in Bangladesh and India. This may be partly due to shorter exposure to contaminated groundwater and a better diet in these countries. The longer-term human health consequences remain to be seen.

Determining the safe level of arsenic in drinking water continues to be a contentious undertaking. In the United States, a practical "safe" level for any contaminant is expressed as the maximum contaminant level (MCL) allowable for public water supplies. MCLs are established by balancing health risks, the technical feasibility for achieving the MCL, and the costs associated with treating the water.

In 1942, the U.S. Public Health Service set an interim drinking-water standard for arsenic of 50 ppb. The U.S. EPA adopted the standard in 1975, but kept it as an interim standard, in large part because of the scientific uncertainties and controversies associated with the chronic toxicity of arsenic.

By 1986, Congress was losing patience with the pace of EPA's standard setting. Amendments to the Safe Drinking Water Act required that the agency finalize its MCL for arsenic within three years, but the EPA had failed to meet this deadline. Meanwhile, studies of arsenic contamination were pointing to the need for a much lower standard. Lowering the MCL much below 50 ppb, however, would greatly increase water treatment costs by utilities in high arsenic regions, such as the Southwest, Upper Midwest, and New England.

Evidence continued to accumulate from epidemiological studies in Taiwan and elsewhere that inorganic arsenic in drinking water could cause several types of cancer, particularly of the lung and bladder. A 1999 report by the National Research Council (NRC) estimated that drinking water with 50 ppb arsenic might induce cancer in as many as one in a hundred people. The NRC minced no words about the arsenic standard, stating that the MCL of 50 ppb "requires downward revision as promptly as possible."[19]

In 2000, the EPA proposed a standard of 5 ppb and requested comments on standards of 3, 10, and 20 ppb. In January 2001, in one of the last acts of the Clinton administration, the EPA proposed an MCL of 10 ppb. Many people thought the standard should be lower, but the EPA concluded that the benefits of the lower level didn't warrant the huge costs of meeting that standard. The newly elected Bush administration delayed adoption of the proposed 10 ppb standard—citing concerns about the science supporting the rule and its estimated cost.

In September 2001, an update of the earlier NRC report concluded that recent studies suggested even greater risks from arsenic in drinking water. This update was released on the eve of the 9/11 attacks on the World Trade Center and Pentagon, yet it still caught people's attention. The following month, the EPA announced that it would adopt the standard of 10 ppb for arsenic. The new standard would take effect in January 2006, giving utilities in places such as Albuquerque, New Mexico, time to adjust. Forty percent of Albuquerque's water supply exceeded the 10 ppb standard. The utility was forced to shut down forty wells and build a $6.3 million treatment plant, the largest microfilter arsenic removal facility in the world.[20]

Considerable debate continues about the risks of drinking water containing low levels of arsenic. The estimated risks are based largely on extrapolating low-dose effects from high-dose (several hundred ppb) studies in Taiwan, Bangladesh, and parts of South America. Some researchers believe that there should be a threshold below which cancer effects do not occur. Others say that such a threshold does not exist. The standards are also based on a lifetime exposure. Compared to people in the United States, villagers in Taiwan are much more likely to spend their lives in one location.[21]

Unlike many other known chemical carcinogens, arsenic appears neither to damage DNA nor cause mutations in genes. Some studies have shown that arsenic can even induce cures in certain forms of cancer. Nonetheless, studies continue to indicate a variety of health risks from exposures to low to moderate levels of arsenic in drinking water, particularly during pregnancy and childhood. Studies now link arsenic in drinking water to a wide variety of adverse health effects, including cancers (bladder, skin, kidney, liver, prostate and lung), vascular and cardiovascular disease, reproductive and developmental problems, cognitive and neurological impairments, immune-system suppression, and diabetes.[22]

Nearly all developing countries, including India and Bangladesh, have retained 50 ppb as the health standard for arsenic in drinking water, despite pressure from the WHO and others to lower the standard to 10 ppb. These countries simply don't have the resources to implement the lower level.

In the United States, the EPA regulates public water supplies to meet drinking-water standards, but it does not regulate private wells. Private well owners are on their own for monitoring the quality and maintaining the safety of their wells. About 14 percent of the U.S. population, or some 43 million people, rely on private wells for their drinking water.[23] The National Ground Water Association and many state governments have programs to educate and assist homeowners with private wells. Notable among these is the state of New Hampshire.

About 46 percent of New Hampshire residents rely on a private well for their water. The vast majority of these wells draw groundwater from crystalline bedrock, a source of high arsenic concentrations in groundwater throughout much of New England and extending into the Canadian Atlantic provinces. Arsenic was considered mostly a local problem in New England until 2003, when the U.S. Geological Survey presented results from regional sampling that estimated more than a hundred thousand people in eastern New England with private wells had arsenic concentrations greater than 10 ppb. About one in five private water wells in New Hampshire contain water above the arsenic MCL.[24]

Researchers at Dartmouth College estimate that 450 cases of cancer could be avoided among the current New Hampshire population if water from private wells was treated to bring arsenic concentrations below the MCL. The study considered only bladder, lung, and nonmelanoma skin cancers. It also did not rely on more recent EPA information (still in draft form at the time) that suggested a much higher risk associated with arsenic in drinking water than previously believed. Hence this is likely an underestimate of the cancer risk and does not account for other health effects.[25]

Maine has a similar arsenic problem. In 2014, a study in Maine grabbed the public's attention when researchers at Columbia University reported that children from three school districts exposed to arsenic in drinking water experienced declines in various measures of IQ and perceptual reasoning. The study assessed 272 children in grades three to five, and its findings were consistent with those of earlier studies in Bangladesh.[26]

Despite the risks, ignorance appears to be bliss when it comes to private wells. A few New Hampshire towns have private well-testing requirements, but most do not. Few households with private wells test them for arsenic. Cost, time, "optimistic bias" (a tendency to think one is safer than one's neighbors), and the absence of any symptoms are all held out as reasons for the lack of testing. "We've been drinking this water for years without any problems," is a common response, or "I had it tested and it's fine," without the slightest understanding of what their well was tested for (many were tested before the arsenic problem was discovered).[27]

Working closely with the U.S. Geological Survey and the Superfund Research Program at Dartmouth College, New Hampshire

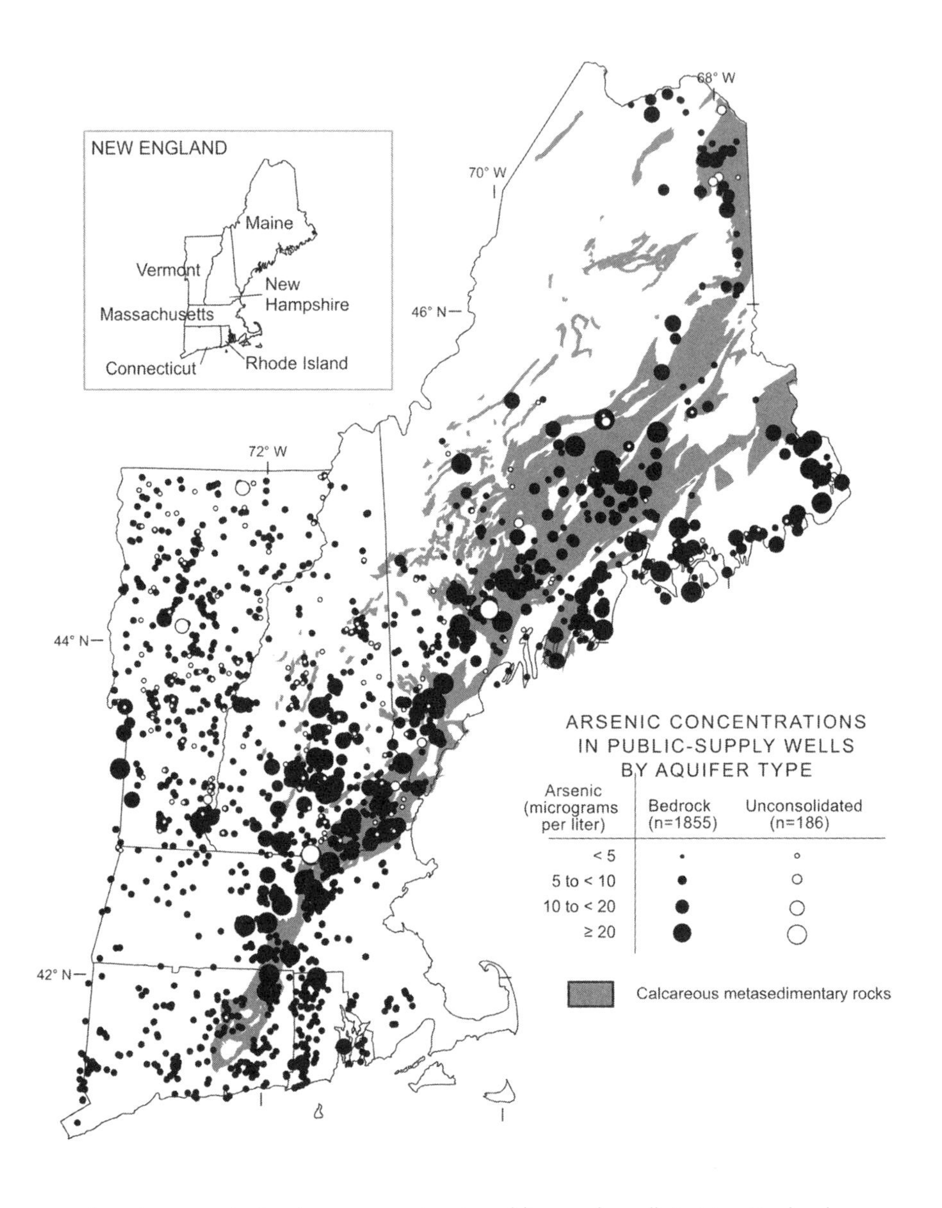

Arsenic concentration in source waters to public supply wells in New England. Note the association of high arsenic concentrations with calcareous metasedimentary rocks. *Source:* Joe Ayotte, U.S. Geological Survey.

environmental and health agencies have worked to raise awareness about the need to test for arsenic, as well as options for treatment. Arsenic is number one on the priority list of hazardous substances at Superfund sites nationwide, and the Dartmouth program examines arsenic levels in both water and rice.[28] Rice, a staple food eaten by half the world's population every day, readily accumulates arsenic from soil—raising yet another health risk factor associated with this contaminant.

New Hampshire agencies have used flyers, websites, social media, and a short film produced by Dartmouth to deliver their message. The agencies also go into the communities where they distribute free test kits during local well-testing days (the owner pays for the lab analysis); participate in town meetings; educate health officers and home inspectors; partner with community-based organizations to intercept people at farmer's markets, churches, and other gathering places; and employ any other means they can think of to get the message out. But that's only the first step.

Many of those who test their wells and find high levels of arsenic still do not take action to treat their water. Installing a water treatment system can be confusing and overwhelming for private well owners. Many people buy the wrong treatment systems or fail to maintain them. To help overcome these problems, the New Hampshire Department of Environmental Services (NHDES) created an interactive web tool, *Be Well Informed,* where a homeowner can enter his or her lab results and get recommendations for treatment.

The NHDES recommends testing when a well is first drilled (or an existing well is deepened), every three to five years thereafter, and at the time of a real-estate transaction. In 2009, the American Academy of Pediatrics issued a statement urging states to require testing of private wells when homes are sold.[29] Proposed laws to address these issues have met with resistance. In New Hampshire, a state whose motto is "Live Free or Die," there's not a lot of support for new regulations. Home builders and realtors are particularly wary of scaring off prospective home buyers with laws that require well testing when homes are sold.

In 2009, the New Hampshire legislature introduced bill HB 1685, which would require testing new wells and wells at homes that were being sold (unless the buyer opted out). The bill died in committee. After years of debate, in 2015, the state legislature passed a less stringent version (HB 498) requiring only that buyers be notified that arsenic (and radon) exists in some well waters and pointing buyers to testing options. These warnings are included (along with innumerable other warnings) in the standard Purchase and Sales Agreement that homebuyers initial and sign.

While much work remains to address the arsenic problem in New Hampshire and elsewhere in New England, dedicated agencies and groups continue working to reduce the risks to homeowners.

Arsenic is one of several naturally occurring contaminants in drinking-water wells. Others include radon (also high in New Hampshire groundwater), radium, uranium, and fluoride. Worldwide, fluoride probably ranks next to arsenic as the naturally occurring contaminant of greatest concern in drinking water.[30]

Fluoride is abundant in nature. In the right quantities, it is essential for the development of teeth and bones. Too much fluoride, however, can lead to dental or skeletal fluorosis. Skeletal fluorosis can be crippling. According to UNESCO, more than 200 million people worldwide rely on drinking water with fluoride levels exceeding the present WHO standard. High fluoride areas extend from Turkey through Iraq, Iran, Afghanistan, India, northern Thailand, and China. India is the most severely affected country worldwide, with more than sixty-six million people, including six million children, estimated to be suffering from fluorosis. The Ethiopian Central Rift Valley is another affected area, where an estimated eight million people are potentially at risk of fluorosis.[31]

Studies of groundwater quality by the U.S. Geological Survey have documented that naturally occurring contaminants are much more likely than human-made contaminants to exceed MCLs (or other human-health benchmarks) in U.S. groundwater. Human activities contribute to these problems in a variety of ways. Irrigation water potentially leaches selenium, salts, and other naturally occurring contaminants. Pumping can draw oxygenated water down into aquifers, potentially releasing arsenic and other contaminants. By mixing waters of different chemistry, managed aquifer recharge can release contaminants such as arsenic from aquifer rocks and sediments into groundwater. Other more subtle human-induced changes in natural groundwater geochemistry can have substantial impacts. Examples from New Jersey and California illustrate these effects.[32]

Sediments of the southern New Jersey coastal plain aquifers do not contain large amounts of radium, yet sampling in the 1990s revealed that radium concentrations in shallow groundwater exceeded the MCL in one-third of wells sampled. Concentrations were particularly high in groundwater beneath agricultural areas. A series of chemical reactions resulting from the application of fertilizers and recharge by septic tanks mobilizes the radium. First, the ammonia and organic nitrogen in the fertilizers and the septic tank effluent is oxidized to nitrate. This nitrification generates acidity, which further lowers the naturally low pH in the aquifer. At low pH, the quartz-rich sediments have little capacity to sorb radium, so it is released into the groundwater. Applying lime with nitrogen fertilizers can further stimulate the release of radium from aquifer sediments into groundwater.[33]

Uranium naturally occurs in much of the eastern Central Valley of California due to erosion of the granitic rocks of the Sierra Nevada. While most people probably think of uranium in terms of its radioactivity, its chemical toxicity to kidneys is the primary concern in drinking water. Human activities mobilize uranium from the sediments to groundwater through a particular geochemical sequence of events. First, irrigation increases plant growth and microbial activity, producing carbon dioxide. The carbon dioxide reacts with mineral solids to form bicarbonate, which then chemically binds with the uranium to keep it in solution. Irrigation and groundwater pumping draw this shallow, uranium-rich groundwater deeper into the aquifer—and potentially toward water-supply wells.[34]

Arsenic, fluoride, radium, uranium, and other naturally occurring contaminants are problematic in many areas worldwide. Developed countries are fortunate in having the means, and multiple ways, to counteract these problems for public water supplies. Wells can be modified to seal off geologic zones that have high levels of naturally occurring contaminants. Well water can be blended (diluted) with water from other sources, or other sources can be used entirely. Finally, treatment can be targeted to meet drinking-water standards. In developing countries and some rural settings, however, providing drinking water that's free from unsafe levels of naturally occurring contaminants remains a major public health challenge.

14

Fracking

Formula for success: rise early, work hard, strike oil.
—J. Paul Getty

Few environmental issues have received more media attention than fracking (hydraulic fracturing). Some see the resultant surge in oil and gas production as a step toward creating jobs and energy independence. Others hold that fracking is wreaking havoc on the environment and human health. These opposing views often are expressed as absolutes. Industry executives have testified before Congress that there are no documented cases of contaminated groundwater resulting from fracking—not a single one.[1] In contrast, anti-fracking activists have claimed that fracking unleashes hundreds of cancer-causing chemicals into the environment and that groundwater supplies have been broadly contaminated. The truth lies somewhere in between.

Fracking uses a fairly simple concept to stimulate oil and gas production. Fluids (mostly water, along with some chemical additives) are injected under high pressure to create fractures in the surrounding rock formations. Sand or another proppant contained within the fluid helps to keep the fractures open. Oil and gas then flow through the fractures and up the production well to the surface.

When petroleum engineers use the term "fracking," they are referring specifically to this process of hydraulic fracturing. To those outside the industry, however, fracking often encompasses all aspects of unconventional oil and gas development. This difference in definitions has contributed to the confusion and controversy.

Hydraulic fracturing is far from new. The technique has been used by the oil and gas industry on vertical wells since the late 1940s. The basic idea of stimulating wells goes back even further, to the beginning of the oil industry in Pennsylvania, when Colonel Edward Roberts patented the technology of launching a torpedo into a well bore. The Roberts Torpedo saw some use until oil gushers in Texas, California, and Oklahoma made it obsolete.[2]

What is new is the combination of hydraulic fracturing with advanced techniques for directional drilling. By using directional drilling techniques, wells can be drilled down to gas-rich shales and other unconventional resources, then turned laterally to follow the hydrocarbon-rich beds as far as a couple of miles. The horizontal length is fracked in stages, starting at the far end and gradually progressing toward the vertical portion of the well. Drilling multiple horizontal wells from a single well pad allows access to as much as a square mile (2.6 square kilometers) of shale located more than a mile (1.6 kilometer) below the surface.[3]

The fracking boom began in the Barnett Shale near Fort Worth, Texas, when George Mitchell, an independent gas producer, figured out the right combination of horizontal drilling, hydraulic pressure, and fracking fluids. After sixteen years of trial and error, while most everyone else was looking overseas for new conventional oil and gas resources, Mitchell's tenacity paid off. Subsequent technological advances opened up vast shale and other "tight" formations to oil and gas development. Shale gas development, which is at the center of much of the controversy, now provides about half of total domestic natural gas production.[4]

A key difference between conventional and unconventional oil and gas wells is how rapidly production in unconventional wells declines within just a few years due to the dispersed, rather than pooled, nature of the resource. At its peak around 2014, tens of thousands of unconventional wells were drilled each year to maintain U.S. production. Using today's techniques, full development of the Marcellus Shale (which underlies much of the northeastern United States) would require more than one hundred thousand horizontal wells, each about 6,500 feet (2 kilometers) long. Each horizontal well costs about $7 to $9 million, with up to twelve wells originating from a single drilling pad.[5]

Shale gas development has been a lightning rod for opposition in the United States, Canada, France, Germany, Poland, South Africa, and the United Kingdom. A primary reason is the sudden appearance of intensive industrial operations, traffic, and almost nonstop noise next to people's homes and in otherwise tranquil rural areas. Accompanying this development is a perceived (or real, depending on your perspective) imbalance of power between ordinary people on one side and "big oil" and "big money" on the other. Other flashpoints include deterioration of local air quality, habitat fragmentation, large use of local water resources for fracking operations, and the disposal of wastewater generated from hydraulic fracturing. And then there's the big surprise that wasn't on anyone's radar until it started happening. Seismic activity in the central United States is on the rise and linked to underground wastewater injection from fracking operations. In 2014, Oklahoma became the "earthquake king" of the lower forty-eight states, exceeding California in the number of earthquakes of magnitude three or greater on the Richter scale.[6]

In addition to these local issues, many people are philosophically opposed to shale gas development because it perpetuates reliance on fossil fuels. Natural gas from shale sources was originally touted as a relatively "green" fuel that could be used during the transition to renewable energy sources. This view has been challenged as greater-than-expected amounts of methane (a potent greenhouse gas) have been found to be leaking from gas operations. Finally, much of the current controversy revolves around the effects on groundwater quality. Images of flaming faucets and exploding well houses have further galvanized public opposition.

Chemicals used in hydraulic fracturing facilitate flow from the shale gas reservoir by reducing friction and inhibiting formation of scale and microbial films that can clog up the works. Chemical additives include acids, surfactants, biocides, corrosion inhibiters, and friction reducers. There are hundreds of different chemicals to choose from, but usually a dozen or fewer are used in any given fracking operation.

Industry spokespeople emphasize that some of these chemicals are common food additives. Be that as it may, other fracking chemicals (such as biocides and corrosion inhibiters) are known to be toxic. Hydraulic fracturing chemicals constitute only a small percentage of the fracking fluid (2 percent or less), but this is of little consolation given that many chemicals are toxic at part-per-billion concentrations. The secrecy surrounding the composition of these fluids has added fuel to the fire.

While fracking below drinking-water aquifers is getting most of the attention, the larger risk is at the land surface from leaks and spills during storage and everyday handling of fracking chemicals. These surface activities are similar to those of other hazardous wastes, with risks managed through regulations and adequate enforcement. Nonetheless, the sudden appearance of the shale gas drilling boom outpaced the ability of some regulatory agencies to keep up with these environmental concerns.[7]

Perhaps the single biggest threat to the environment is when water comes back to the surface, where it's impounded for disposal, treatment, or reuse. This is no piddling amount. During the first weeks of a hydraulic fracturing operation, the amount of water often exceeds one million gallons, then diminishes over time. This wastewater is a mixture of injected fracking fluids and water from the shale formation. Most shales were deposited in marine environments and have high salinities, along with high levels of naturally occurring elements such as barium, strontium, and radium. Disposing of this water is a major issue. Most of the wastewater is injected underground into deep wells, which accounts for the spike in earthquakes.

Another concern falls in the low-risk, high-impact category. It's possible that fracturing fluids and deep brines could migrate upward to overlying aquifers through fractures created by hydraulic fracturing that connect to natural faults. Available evidence indicates this risk is very low. Several thousand feet of rocks, low upward hydraulic gradients, and the high density of fluids generally serve as a natural barrier to such transport. In addition, oil and gas companies design their fracking operations to maintain fractures within the targeted shale gas zone. Fracking is extremely expensive, and it's a waste of money to fracture outside the production zone. Perhaps the main irony of the fracking controversies is that the act of hydraulic fracturing itself probably poses the lowest risk. This specific low risk allows oil and gas executives to claim that there are no documented cases of groundwater contamination caused by "hydraulic fracturing."

Even so, there are possible exceptions to this low-risk scenario.[8] The first possibility is in geologic settings where large thicknesses of low-permeability rocks do not separate the fracked zone from freshwater aquifers. The second exception is where the pressure pulse from hydraulic fracturing could affect nearby production wells, potentially leading to well failure and release of fluids. A third scenario is where abandoned oil or gas wells that have not been properly sealed could provide a fast track for contaminants to reach freshwater aquifers. States now have strict rules for sealing abandoned wells, but this hasn't always been the case. For decades, many oil and gas wells were just left open or were "creatively" abandoned with logs, brush, or any other debris that was lying around. Hundreds of thousands of abandoned oil and gas wells fall in this category. Time passes, vegetation recovers, and many of these wells have disappeared from view. States have "orphan well" programs to try to locate and properly plug these wells, but it's easier said than done and a long way from completion.[9]

The most likely avenue for shallow aquifer contamination from below the land surface is through migration of stray gas (primarily methane) along the outside of well casings. This can occur along defects caused by factors such as shrinking cement or improper grouting between the well casing and the surrounding rocks. The gas could originate from the production zone or (perhaps more likely) from intermediate gas-bearing zones penetrated by the well.[10]

Several layers of steel pipe casing and cement shield the aquifers from contaminants inside the well. The cement filling the space (annulus) between the outermost well casing and the rocks is particularly important to the well's integrity. To be effective, the cement must isolate the right zones, be of high quality, and bond robustly with the casing and outside rock formations. If any of these requirements aren't met, stray gas can be a problem. Cement also degrades with time (as can well casings), creating the potential for future contamination.

Stray gas leaking along the outside of well casings is an old and stubborn problem that is very familiar to the oil and gas industry. Estimates of the frequency of leaking wells range from less than 1 percent to 9 percent or more. Due to its buoyancy, methane gas can travel quickly up along defects in the cemented annulus. By contrast, dense saline groundwater moving along the annulus and into freshwater aquifers is less likely, and would take a lot longer.[11]

Methane by itself is not considered a health hazard in drinking water, but it can lead to other water-quality problems. Bacterial oxidization of methane can increase the solubility of arsenic, manganese, and iron. Anaerobic bacteria can convert dissolved sulfates into sulfides, causing odors and decreased well yields. Methane gas also can resuspend fine-grained sediments that accumulate at the bottom of wells, resulting in cloudy, bubbly water. In rare cases, an accumulation of methane gas can result in an explosion.[12]

The basic question of under what circumstances methane gas will cause significant harm to groundwater remains unresolved. Because methane is not considered a drinking-water contaminant, it was rarely analyzed in water-well samples prior to the shale gas boom. As a result, few baseline data exist. Compounding matters, many water wells have naturally occurring methane. Lighting water from faucets in Pennsylvania is a party trick that began long before shale gas development.

Sampling groundwater to assess the effects of shale gas development has been almost exclusively limited to household and farm wells. Most of these data have not been collected for scientific studies, but rather to address liability concerns. Domestic wells are not designed for collecting representative samples of the groundwater system (areally or with depth) and are poor instruments for studying contaminant processes and impacts. An additional problem is that many claims of contaminated domestic wells are settled off the record, which means their water quality has not been documented. How many of these claims are legitimate is unknown.[13]

Not surprisingly, studies to date have resulted in contradictory findings and a lack of consensus within the scientific community. The best-known example comes from northeastern Pennsylvania. In 2011, Duke University researchers reported that concentrations of methane and other components of natural gas in well water were higher for homes within 0.6 miles (one kilometer) of drilling sites, implicating oil and gas development. Two years later, a separate study found that methane concentrations in water wells primarily correlate with topographic features such as valley bottoms, and not proximity to gas wells. Two years later, another team of researchers investigating the same area stated their main conclusion in the title of their paper, "Methane Concentrations in Water Wells Unrelated to Proximity to Existing Oil and Gas Wells in Northeastern Pennsylvania." More recently, a team (comprised in part by the original Duke University researchers) used noble gas isotopes to demonstrate gas leakage along well annuli and faulty casings at selected homeowner wells that are near gas drilling sites and have high methane concentrations. All of these studies were published in well-respected, peer-reviewed journals.[14]

Insufficient baseline data and conflicting reports have stymied hydrogeologists in developing any firm estimate of the extent of the stray gas problem. Much more rigorous groundwater monitoring in specially designed wells is needed to achieve a reasonable understanding and consensus on existing impacts—and to develop better predictions of long-term effects. Sampling needs to be conducted closer to well pads, at multiple depths, and over a longer timeframe.[15]

Anti-fracking activists and the oil and gas industry have both contributed to this lack of research. Anti-fracking activists are convinced that fracking is dangerous, and simply want to ban it. The oil and gas industry fears that such studies will only bring bad news, yet industry involvement is essential in order to gain access to sites for research. Finally, federal funding to support such research has been noticeably absent. Caught somewhere in the middle are the basic, and largely unanswered, environmental questions related to this huge and controversial industry.

15

Nitrate and Aquifer Protection

Man is a complex being who makes deserts bloom and lakes die.
—G. B. Stern

Susan Seacrest was a very worried mother. From the time her son, Logan, was an infant he had been frequently hospitalized for serious digestive difficulties. Among the possible causes, at least she didn't have to worry about living in an unhealthy environment. One of the perks of Nebraska farm country was all that fresh air and clean water. After four years of terrible anxiety, Logan mysteriously recovered.[1]

Soon thereafter, Seacrest read a newspaper article describing potential environmental health hazards right in her own backyard. A Nebraska doctor, concerned about what appeared to be elevated levels of leukemia and non-Hodgkins lymphoma in the Central Platte River Basin, had convinced the University of Nebraska Medical Center to hire an epidemiologist. Dennis Weisenberger was now investigating potential links between cancer and factors such as pesticide and fertilizer use and groundwater contamination.

Seacrest wrote to him, wanting to know about his research, any possible theories he might be developing, and how she could protect her children. Weisenberger responded by telling her to learn everything she could about groundwater, and then get actively involved. She accepted the challenge. Housework went on the back burner and her kitchen was soon piled with articles, studies, and books. She talked to public officials and anyone who knew anything about groundwater.

Then she began to share her knowledge and concerns with those who showed the slightest interest. Only partly in jest, her husband made a sign for her to wear at parties: "Do NOT ask this woman about groundwater."

Susan Seacrest's interest in groundwater moved beyond protecting the health of her own family. In 1985, she founded the Groundwater Foundation, a nonpolitical initiative to educate people about groundwater. She soon ran up against a wall of skepticism. People told her no one was going to pay attention to education—it would take regulation to get people to change. But she kept going and attracted a core group of volunteers.[2]

In 1988, the group launched the Children's Groundwater Festival. Once again, people were skeptical that anyone would care enough to attend an event about water. Their expectations were modest—maybe fifty kids would sign up. Over 1,700 registered for the event. Since that first Children's Groundwater Festival, the concept has been replicated in all fifty states and fifteen foreign countries.[3]

In 1994, the Groundwater Foundation launched another program, Groundwater Guardian, as a way to open dialogue and generate support among communities that were actively involved in protecting their groundwater. The foundation now works with hundreds of communities in North America that are focused on source-water protection, proper well-abandonment programs, and public education. In addition, the Groundwater Foundation has worked with farmers in Nebraska and other agricultural states, educating them about the need to reduce nitrate levels on their crops.

Susan Seacrest has served on the EPA's Children's Health Protection Advisory Committee and the National Drinking Water Advisory Council. She has been a keynote speaker at the United Nations, as well as other conferences around the world. In 2007, this Nebraska mother, who educated herself about groundwater and "got actively involved," received the prestigious Heinz Award for the Environment.

Nitrate is the world's most widespread groundwater contaminant, endangering human health and the environment. Infants who drink water with high levels of nitrate (often mixed with baby formula) can become seriously ill from "blue-baby syndrome," a condition in which an infant's red blood cells cannot carry enough oxygen. If not properly treated, blue-baby syndrome can be fatal. Excessive nitrate levels also have been linked to cancer, reproductive damages, and respiratory and thyroid problems, but these connections are more tenuous.

There is, however, no doubt about the environmental effects. Nitrate from groundwater discharging to surface-water bodies contributes to excessive growth of algae and other nuisance aquatic plants (eutrophication). In addition to interfering with recreational activities such as fishing, swimming, and boating, decay of algae can result in large dead zones from lack of oxygen. This situation, known as hypoxia, is a growing menace to surface water. A dead zone in the Gulf of Mexico that has developed from nutrients discharged by the Mississippi River periodically grows larger than the state of Connecticut.[4]

Nitrate is part of a global nitrogen cycle in which nitrogen circulates among the atmosphere, soil, and water. Although nitrogen is essential for plant and animal life, most people are barely aware of its existence, much less that it is so abundant. We tend to think the air we breathe is oxygen, but in fact it is almost 80 percent nitrogen gas. With every living organism on the planet being critically dependent on nitrogen, how has nitrate become such a major environmental hazard?

The answer begins with chemistry. Held together by extremely tight chemical bonds, atmospheric nitrogen is unusable by most plants without some sort of intervention. Help comes from certain bacteria and algae in the soil and water, which "fix" nitrogen by changing it into a form usable by plants. Nitrogen-fixing bacteria also live in the roots of legumes, such as peas, clover, and alfalfa, where they have a symbiotic relationship with their host. The plants supply the bacteria with food. In return, the bacteria secrete ammonium compounds (a form of nitrogen) that are absorbed by the legumes and other plants growing in the same soil. The relatively immobile ammonium ion is readily converted to nitrate through a process known as nitrification.

Despite nature's help, a scarcity of nitrogen in usable form, as nitrate or ammonium, often limits plant growth. The turning point came in the early 1900s when two German chemists, Fritz Haber and Carl Bosch, developed and commercialized a process to break the chemical bonds in atmospheric nitrogen and combine the freed nitrogen atoms with hydrogen to form ammonia. Haber and Bosch received Nobel prizes in chemistry for their work, but their primary intentions had nothing to do with helping plants grow. The ammonia they produced was largely used for explosives and mustard gas. It wasn't until after World War II that ammonia production began to be focused on fertilizers. Today, 120 million tons of nitrogen are extracted annually from the air to produce ammonia for fertilizer. This energy-intensive process consumes about 2 percent of the world's energy.[5]

The synthesis of ammonia is considered one of the most significant technological innovations of all time. Without today's large-scale use of nitrogen in agriculture, as much as 40 percent of the world's population would starve without major changes in global food consumption. On the downside, the use of fertilizers (and other sources of nitrate, such as animal waste and septic tanks) has led to widespread nitrate contamination.[6]

The high loading of nitrate into soils and groundwater has been taking place for a relatively short period of time compared to groundwater travel times. This makes it mostly a shallow groundwater problem—for now. As nitrate moves downward, groundwater quality in many aquifers is on a long-term decline.[7]

How far nitrate will penetrate into groundwater, and whether it makes its way to surface water, largely depends on denitrification—a process that converts nitrate into nitrogen gas and so returns the nitrogen to the atmosphere. Denitrification occurs after bacteria have used up the oxygen in groundwater. Wetlands and poorly drained soils are ideally suited to denitrification, demonstrating one of several ways in which wetlands are highly beneficial to the environment. Denitrification also occurs at the interface of groundwater with streams and lakes, as water passes through organic-rich sediments low in dissolved oxygen. Understanding where denitrification is taking place is key to

Urban encroachment may result in public supply wells withdrawing groundwater that was recharged through agricultural lands. Photograph by Lynn Betts, U.S. Department of Agriculture, Natural Resources Conservation Service.

determining the vulnerability of groundwater and surface water to nitrate contamination.

Despite widespread awareness of this growing health and environmental problem, little progress has been made in reducing nitrate contamination of groundwater. With global food, feed, and fiber demands anticipated to increase by more than 70 percent in the next several decades, the problem will only intensify.[8] And then there's the rise of biofuels production. Corn, the primary biofuel source in the United States, requires more nitrogen fertilizer per acre than almost any other crop.

The U.S. National Academy of Engineering regards managing the nitrogen cycle as one of fourteen "grand challenges" for engineering in the twenty-first century.[9] Even if nitrate loading into groundwater miraculously stopped today, the problems would continue for decades. Consider the Chesapeake Bay.

The Chesapeake Bay is the nation's largest estuary and home to a wide variety of fish, wildlife, and plants. Renowned for its striped bass, oysters, and blue crabs, only the seafood harvest of the Atlantic and Pacific Oceans exceeds that from the Chesapeake Bay. The bay is also the winter home for about one million waterfowl, due to its central location along the Atlantic Flyway.

For decades the Chesapeake has suffered from nutrient overload, particularly from animal waste and chemical fertilizers. These excess nutrients lead to a familiar sequence of events. The nutrients stimulate algal blooms. The algal blooms decompose to cause large areas of low-dissolved-oxygen concentration. The low oxygen concentrations kill aquatic life, with crabs and other bottom-dwelling organisms being particularly vulnerable. The algal blooms also block the sunlight needed by submerged grasses. When those grasses die, an important food for waterfowl—as well as shelter for crabs and juvenile fish—is removed.

In 1983, the governors of Maryland, Virginia, and Pennsylvania, the mayor of the District of Columbia, and the EPA administrator signed an agreement leading to the Chesapeake Bay Program. After decades of studies with little follow through, officials finally were taking actions to reduce nutrients and improve water quality in the bay. Progress has been slow, and further complicated by the eventual recognition that about half of the nitrogen load to the Chesapeake Bay is coming from groundwater.[10] Groundwater discharges along the shoreline, but mostly enters as baseflow to streams and rivers that flow into the bay.

The states and the EPA have set specific goals to reduce nitrogen loads to the bay. These depend in large part on implementation of best management practices (BMPs) by farmers in the watershed. Meeting the goals, however, is complicated by the time it takes for water to move through the groundwater system. One study shows that if nitrogen inputs on the eastern shore of the Chesapeake Bay could immediately be reduced by 40 percent at the water table, it would still take about forty years to realize the program's goals. This multi-decade lag between program implementation and achieving the desired reduction of nitrogen loads to the bay makes it challenging to evaluate the effectiveness of the BMPs. It also makes it difficult to explain the slow progress to farmers who have been asked to modify their practices in order to "Save the Bay."[11]

In addition to the effects of high nitrate concentrations on ecosystems such as the Chesapeake Bay, many agricultural areas are dealing with the human health consequences. In 2012, the University of California, Davis released a comprehensive evaluation of nitrate contamination in the Tulare Lake Basin and Salinas Valley, California. Located in the southern San Joaquin Valley and southeast of Monterey, respectively, these areas account for about 40 percent of California's irrigated cropland and over half of the state's dairy herd. The study area, containing four of the five counties with the largest agricultural production in the United States, is also one of California's poorest regions.[12]

As in most agricultural regions, fertilizers and animal wastes applied to cropland are by far the largest sources of nitrate in the region's groundwater. About 2.6 million people rely on this groundwater for their drinking-water supply. Most of this population (typically the wealthier segment) is protected by systems that treat the water or blend it with cleaner water to reduce nitrate concentrations. However, about 10 percent of the population (about a quarter million people) relies on domestic wells and small systems that don't receive such treatments. In addition, nitrate problems often are compounded by naturally occurring arsenic, chromium, uranium, and other groundwater contaminants. In general, it is the small Latino farmworker communities whose drinking water exceeds the EPA nitrate standard. These communities do not have the tax base to support construction and operation of treatment plants, or to secure alternative sources of water. As a result, many poor people spend 5 to 10 percent of their income to purchase drinking water, filling up their five-gallon jugs at local stores or water vending machines.[13]

In 2012, California enacted law AB 685, thereby becoming the first state to recognize the human right to safe, affordable water without discrimination. Although this law looks good on paper, state agencies are only required to "consider" this policy when they adopt or revise regulations. They are not required to provide clean water or allocate additional resources to fix subpar water systems. Current efforts to reduce nitrate loads to groundwater won't have any real impact for decades, which means these communities urgently need safe and affordable water.[14]

On May 24, 1610, as part of a civil code for Jamestown, Sir Thomas Gates, lieutenant governor of Virginia, established the first wellhead protection area in the United States: "There shall be no man or woman dare to wash any unclean linen . . . nor rinse or make clean any kettle, pot or pan, or any suchlike vessel within twenty feet of the old well or new pump. Nor shall anyone aforesaid within less than a quarter mile of the fort, dare to do the necessities of nature, since [by] these unmanly, slothful, and loathsome immodesties, the whole fort may be choked and poisoned."[15]

Nearly four hundred years later, the 1974 Safe Drinking Water Act omitted this crucial concept of source water protection and focused almost exclusively on the water coming out of taps. It gradually became obvious that it's much more effective to protect water at its source than to rely solely on expensive treatment at the delivery end. In 1986, amendments to the Safe Drinking Water Act required states to evaluate the sources of their drinking water and the factors that threaten these sources. States have completed source water assessments for virtually every public water system, from major metropolitan areas to the smallest towns. Even schools, restaurants, and other public facilities with wells were assessed. These assessments, however, have not been kept up-to-date. A group of federal agencies and organizations known as the Source Water Collaborative is calling for a recommitment to assessing and protecting drinking-water sources.[16]

Denmark has what's considered the world's most exemplary aquifer protection program. This is not surprising, given that all of the country's drinking water comes from groundwater. Adding further impetus, about two-thirds of Denmark's land area is devoted to agriculture. Concerns about fertilizer and pesticide use top the list, but there's also the challenge of managing the country's booming livestock industry. In a country of about 5.6 million people, some twenty-five million pigs are raised each year, presenting a major manure challenge. The country's flat glacial outwash plains and undulating hills provide a plentiful and easily accessible groundwater resource, but also make it highly vulnerable to contamination.[17]

Cholera pandemics in Europe in the 1800s, including a severe outbreak in Copenhagen in 1853, were instrumental in Denmark's turn to groundwater. In addition, surface-water resources are limited. Streams are mainly groundwater-fed and relatively small.

The Danish government's official position is that drinking water must be based on groundwater that needs only simple treatment (aeration, filtration, and pH adjustment) before being delivered to the populace. Except for Copenhagen, drinking water is not even chlorinated. With farming-related contamination increasingly threatening their groundwater, Denmark has developed a state-of-the-art source water protection program.

A national network for monitoring groundwater levels began in 1951, followed by monitoring of water quality in the late 1980s. These data, collected over many years, provided a solid foundation when the Danish government launched a ten-point initiative to improve groundwater protection in 1995. All of Denmark was classified into three types of areas based on their value for groundwater abstraction—particularly valuable, valuable, and less valuable. No area was considered not valuable.

Within areas designated as particularly valuable (about 40 percent of the country), the Danes undertook an ambitious hydrogeological mapping program to produce an accurate picture of aquifers and their vulnerability to contamination, especially by nitrates. The program is remarkable in how it combines monitoring, borehole information, geologic mapping, and advanced airborne geophysical methods with a sophisticated set of modeling tools.[18]

From the outset of their mapping program, the Danes realized that borehole data provided only a very coarse picture of the nation's aquifers. For example, these data are limited in mapping buried glacial valleys that may be important to groundwater flow. To supplement the ground-based data, Danish researchers developed new airborne geophysical mapping methods. Low-flying helicopters systematically fly back and forth, while towing a large wire-loop instrument hanging from a cable. The system transmits an electromagnetic signal toward the ground, then translates the signal it receives back into a three-dimensional picture of resistivity of earth materials to a depth of about a thousand feet (three hundred meters). Scientists analyze these data much like a doctor examines brain scans to create a three-dimensional picture of subsurface geology. Existing borehole data are essential for calibrating the technique. New boreholes are drilled in strategic locations to test how reliably the method represents the actual geology. All geophysical data and interpretations are stored in a national database, which anyone can access free of charge. With considerable foresight, Denmark established a national calibration site (a well-characterized piece of land) in order to continually test the techniques as they evolve.[19]

The national mapping program cost about 250 million euros (325 million dollars), which was paid for by water customers through a tax on their water use. Completed in 2015, the maps provide key information useful for establishing site-specific groundwater protection zones and regulating land use to prevent groundwater contamination. The maps are expected to influence future urban development so as to protect the recharge areas to water-supply wells.

These examples, as well as those in previous chapters, illustrate the vulnerability of groundwater to contamination by many pathways—downward percolation from land-surface sources, direct entry of contaminants through wells, pumping-induced intrusion of saline or other poor-quality water, and human activities that mobilize arsenic and other naturally occurring contaminants. Characteristics such as soil type and depth to the water table make some groundwater systems more vulnerable to contamination than others. Yet as a National

Research Council committee once emphasized, "all groundwater is vulnerable."[20] Therefore all drinking-water supplies require vigilance to maintain a safe water supply.

Adding to these challenges, new contaminants continue to emerge. As we are completing this book, the lead poisoning in Flint, Michigan, is making daily headline news. Although unrelated to groundwater, this crisis illustrates the potential consequences for the well-being of an entire city of failing to make safe drinking water a priority. Lead also can be a problem in water from private wells, particularly those with corrosive water that is acidic or contains high levels of salt. In statewide studies, researchers at Penn State and Virginia Tech found that 12 to 19 percent of sampled private wells exceeded the EPA lead action level of 15 ppb.[21] These private wells are not regulated by the EPA and are the homeowners' responsibility.

16

Transboundary Aquifers

It's better to have problems with water than problems without water.
—California governor Edmund G. "Pat" Brown

Memphis, Tennessee, is a music mecca where Elvis, B. B. King, Johnny Cash, and many others got their start. The city sits along the Mississippi River, yet relies completely on groundwater for its public water supply. The use of groundwater dates back to the late nineteenth century when Memphis was a filthy city that invited contagion. After major yellow fever epidemics in the 1870s, the city decided to improve its sanitation and water supply. As part of this transition, it turned to groundwater.

The first use of groundwater occurred in 1886, when a local ice company drilled an artesian well into what would become known as the Memphis Sand. Word spread quickly and crowds collected around the well. A city engineer described the jubilant scene: "The elixir of life had been found. Memphians of all degrees, high and low, old and young, with buckets and jugs, coffeepots and tin cans, waited in long files to be served, each in turn, from the gushing, hygienic well. And so for days. In good weather there could be seen lines of baby carriages, each with its little occupant . . . Physicians gave prescriptions: 'Let the baby drink artesian water.'"[1]

The Memphis Sand proved to be a prolific source of high-quality groundwater. Pumping increased as it became the city's sole water supply, but not everyone was happy with this development. Memphis lies just north of the Tennessee-Mississippi border, and some of the city's wells lie almost on the state line. In recent decades, Mississippi has become increasingly concerned about Memphis sucking groundwater out of their state. Mississippi officials claim that Memphis is the biggest user of its groundwater. Mississippi relies heavily on groundwater. It ranks among the top ten states pumping groundwater—exceeding even Arizona.[2]

Like all states, Mississippi views groundwater lying within its borders as a sovereign resource. In 2005, the state sued Memphis and its water utility in district court, seeking over $1 billion in damages. The court ruled that the case was an interstate issue and so must be heard by the U.S. Supreme Court. In 2010, the Supreme Court dismissed the case "without prejudice," meaning that Mississippi could refile if it could quantify damages. In 2014, Mississippi was back before the high court, arguing that Memphis had "forcibly" taken 252 billion gallons (950 billion liters) of Mississippi's groundwater since 1985. In 2015, the Supreme Court appointed a special master to hear the case.[3]

Mississippi's ongoing legal battle with Memphis over groundwater rights illustrates the need for interstate agreements to manage certain transboundary aquifers. Currently, no such agreements exist. At one point officials from Tennessee, Mississippi, and Arkansas had discussed developing a plan to jointly manage the Memphis Sand Aquifer. Mississippi's group, frustrated with the slow pace of the negotiations, withdrew from the discussions.

The challenges for sharing groundwater resources are magnified for aquifers crossing international borders. Several hundred such transboundary aquifers have been identified, underlying almost every non-island country in the world.[4] While hundreds of treaties have been completed for international river basins, only a handful of modest agreements exist for transboundary aquifers.

The most authoritative statement on the law of shared groundwater resources is the U.N. International Law Commission's 2008 Draft Articles on the Law of Transboundary Aquifers. This "law" consists of nonbinding articles that the U.N. General Assembly simply suggests that countries "take note" of. The draft articles include references to protecting recharge and discharge zones, ensuring the functioning of aquifers, and monitoring. Basic principles include equitable and reasonable use, the obligation not to cause significant harm, and the duty to negotiate.

Among the hundreds of transboundary aquifers, only two agreements provide for any joint management. The longest-running management agreement concerns the Genevese Aquifer, a small aquifer shared by Switzerland and France near Lake Geneva. The agreement, signed by the Swiss Canton of Geneva and the French prefect of Haute-Savoie, centers on sharing the costs of artificial recharge and controlling groundwater abstractions. First signed in 1978, the agreement was extended for another thirty years in 2008—a modest but true success story.[5]

In 2015, Jordan and Saudi Arabia entered into an agreement for the management and use of part of a large nonrenewable transboundary aquifer known as the Al-Sag in Saudi Arabia and the Al-Disi in Jordan. Saudi Arabia has exploited the Al-Sag Aquifer as part of its ill-fated effort to become self-sufficient in wheat. Jordan uses the aquifer for various purposes, including piping water two hundred miles (325 kilometers) to Amman for its public supply. Both countries are rapidly depleting the aquifer. Their agreement includes a protected area along the border where all activities that depend on groundwater must be discontinued within five years, and a management area where groundwater use is restricted to municipal supply.[6] Time will tell how the agreement plays out between these two water-strapped countries.

The Guarani Aquifer, which underlies Argentina, Brazil, Paraguay, and Uruguay, is another transboundary aquifer held up as an example of international cooperation. The Guarani Aquifer is one of the world's largest aquifer systems, covering an area bigger than Texas and California combined. In 2010, the presidents of the four countries signed a joint agreement pledging their cooperation toward sustainable management of the aquifer. There were no major crises to resolve, but demands on the aquifer were growing for agriculture, drinking water, and geothermal energy. This bare-bones agreement simply emphasizes each country's sovereign rights and the need to share data and information. Nonetheless, it's a promising start.[7]

The Abbotsford-Sumas Aquifer, shared by Canada and the United States, is an example of cross-border groundwater contamination. Nitrates from intensive agriculture (mostly blueberries and raspberries) migrate from the province of British Columbia to Washington State in the shallow sand and gravel aquifer. Efforts by a binational group of stakeholders, with the blessing of federal, state, and provincial governments, have led to changes in agricultural practices, as well as programs aimed at raising grower awareness. Despite these efforts, little change in nitrate concentrations has occurred, illustrating once again the difficulty of remediating groundwater contamination.[8]

Probably the best-known, and most contentious, transboundary aquifer is the Mountain Aquifer that underlies Israel and the West Bank. The aquifer serves as the sole source of water for the West Bank, and is a key component of Israel's water security. The Mountain Aquifer suffers from serious groundwater depletion, contamination, and saltwater intrusion. Negotiations on water use between the Israelis and Palestinians have waxed and waned with the peace process, and yet have been surprisingly resilient when all else has broken down. Little concrete progress has been made, yet the Mountain Aquifer exemplifies the common overstatement that this century's wars will be fought largely over water. Water issues may be part of disputes and exacerbate political and social tensions, but as an essential resource, water has more often been a focus of negotiations.[9]

While few agreements address specific transboundary aquifers, groundwater is increasingly mentioned as part of international water agreements. Between 2000 and 2007, more than half of all international water agreements had some provision for considering groundwater.[10] A recent effort along the U.S.-Mexico border illustrates opportunities and challenges in addressing transboundary aquifers.

The U.S.-Mexico border runs nearly two thousand miles (3,200 kilometers) from the Pacific Ocean to the Gulf of Mexico. In the late 1960s, the border population began to grow rapidly as maquiladoras (manufacturing plants that take advantage of duty-free parts and cheap labor) set up shop along the Mexican side of the border. This growth accelerated after the North American Free Trade Agreement (NAFTA) was put into effect in 1994. By 2010, more than fourteen million people were living along the border, with an anticipated population increase of 40 percent by 2020.[11]

Transboundary aquifers are an essential, and in many cases, the sole source of water for communities on both sides of this hot and dry border. This critical resource has been overexploited and polluted by untreated wastes, agricultural chemicals, and industrial byproducts. Despite its importance and serious problems, groundwater along the border has been largely ignored by both the United States and Mexico.[12]

There are several reasons for such lack of interest. Groundwater management ranks relatively low in priority compared to illegal immigration, drug trafficking, violence, and economic trade. Considerable complications also arise because of differences in how groundwater is managed on the two sides of the border. In Mexico, water belongs to the nation and water management is centralized within the federal government. In the United States, federal environmental laws apply to water quality, but the allocation of water resources falls within the almost exclusive purview of each individual state. States are extremely wary of any federal intervention into their water management role. Developing a unified U.S. position on transboundary aquifers is complicated by the starkly different ways in which groundwater is managed in each border state.

In 2006, the U.S. government enacted Public Law 109–448 to "establish a United States-Mexico transboundary aquifer assessment program (TAAP) to systematically assess priority transboundary aquifers." The U.S. Geological Survey was responsible for implementing the act in partnership with the water resources research institutes in Arizona, New Mexico, and Texas. Funding would be shared equally between the USGS and the state institutes.

The "priority aquifers" are carefully defined by the act. Four aquifers were identified as first priority. Two of these, the Santa Cruz and San Pedro aquifers, are shared by Arizona and the Mexican state of Sonora. The other two, the Hueco Bolson and Mesilla Basin (named the Conejos-Médanos Aquifer system in Mexico), are shared by New

Mexico, Texas, and the Mexican state of Chihuahua near the border cities of El Paso and Juárez. Additional priority aquifers under the act were authorized only for New Mexico and Texas, not Arizona. This restriction avoided selecting aquifers in the Colorado River Basin, which was already the focus of international agreements (and disagreements).[13]

California was noticeably missing from the act. The state opted out, in large part to avoid creating a forum for Mexico to object to California lining a twenty-three-mile (thirty-seven-kilometer) section of the All-American Canal along the California-Mexico border. Most of the canal leakage (that would be "saved" by lining the canal) flowed to Mexico in the form of groundwater, where it sustained wetlands and farming. The wetlands are a unique habitat for more than one hundred bird species, and had become a sensitive flashpoint.[14]

Mexico had not been consulted in formulating the act and specifying its priority aquifers. Nor had there been any consultation with the International Boundary and Water Commission (IBWC)—the agency through which U.S.-Mexico water issues are discussed and resolved. The IBWC has a U.S. section that is part of the State Department, and a Mexican section that is part of the Mexican foreign ministry. The act instructed that the implementers of the TAAP consult with IBWC "as appropriate."

Part of the reason the act envisioned a minimal role for the IBWC was its lack of experience with groundwater issues. The commission's primary mandate centered on the treaty that allocated water for the United States and Mexico from the Colorado River and Rio Grande. As it turned out, the IBWC would emerge as an essential player in gaining Mexican participation.

Not having been consulted, the Mexicans refused to recognize the TAAP as a U.S.-Mexico program. In their view, it was the "Bingaman Act" (named after New Mexico Senator Jeff Bingaman, its principal architect). Without Mexican support, coordination of activities across the border would be extremely difficult, if not impossible, and data and information from the program would not be officially sanctioned as legitimate by Mexico.

Transforming the program into a true binational effort required building a partnership that respected the legal and institutional frameworks of both countries. These efforts took place at both local and national levels. Technical meetings, site visits, and joint planning activities were conducted to share information on data and modeling, as well as to build support and common understanding of priority needs.

Negotiations involved arriving at some kind of consensus among the dizzying array of participants—the Mexican and U.S. sections of the IBWC, CONAGUA (Mexico's national water commission), the U.S. Geological Survey, and university participants from Arizona, New Mexico, and Texas. Despite the challenges, the two countries made steady progress. On August 19, 2009, a Joint Cooperative Process was signed in a ceremony on the International Bridge between El Paso and Juárez. This agreement established the basic framework for binational aquifer assessment activities, with the IBWC as the official repository for the data and reports.

Over a three-year period, the Mexican and U.S. parties had established the foundation for a genuinely collaborative effort. After successfully working through the myriad challenges of establishing a joint program, funding was cut off in 2011. Although some funding was reinstated in 2016 (the final year of the ten-year authorization period), the program's future remains uncertain. Its story demonstrates how efforts to address transboundary aquifer issues require multiple levels of engagement, mutual respect, and tenacity.

17

Sharing the Common Pool

What's the use of having developed a science well enough to make predictions if, in the end, all we're willing to do is stand around and wait for them to come true?
—F. Sherwood Rowland

Many people perceive groundwater as a privately owned resource and resist any governmental role in guarding it. A cattle rancher near Wilcox, Arizona, an area where people's wells are going dry due to overpumping, sees it this way: "I might wake up tomorrow and decide I want to start farming. I have that God-given right. If someone has the money and the initiative to go out and farm, if they can drill to 10,000 feet deep and build [grow] a banana tree on top of the mountain, if they can make it work, then more power to them. That's what makes this country so great!"[1]

While this is an extreme statement, the attitude that "it's my land and I'll pump as much as I want" is all too pervasive. This view regards groundwater as a static resource no different than a mineral or oil deposit underlying one's land. It also clings to past values that no longer make sense in today's world. Virtually all of the world's most pressing environmental problems require cooperation and collective action. Groundwater depletion and contamination are prime examples.

The challenges of getting people to recognize the seriousness of groundwater problems are not unlike those associated with climate change. It's hard to understand the scope of a problem you can't see and that develops slowly. These challenges are further exacerbated by the human propensity to deny a problem for as long as possible.

Every groundwater situation poses unique challenges. At one extreme is mining fossil water from aquifers having no, or minimal, present-day recharge. Such aquifers include the southern High Plains in the United States, many aquifers in North Africa and the Middle East, and the Kalahari Sands in southern Africa. For these systems, the critical questions are how much groundwater is economically recoverable, what are the appropriate rates of depletion, and how society can orchestrate a "smooth landing" from excessive groundwater dependence. These questions are rarely addressed in any serious manner. Instead, a race to the bottom is under way in many of these aquifers.

At the other end of the spectrum are aquifers in geologic and climatic settings where groundwater pumping can be maintained indefinitely. Withdrawals during dry periods are balanced by replenishment during intervening wet periods. Many of these aquifers, however, are shallow and particularly vulnerable to contamination. In these situations, protecting groundwater quality is of paramount concern.

Many, if not most, groundwater systems fall somewhere between these two situations. The task of managing these intermediate aquifers is complicated by the long response times of the groundwater flow systems—another challenging concept for many people to grasp. Yet perhaps the least understood concept is that only a small part of the total amount of groundwater in storage may be available for use without unacceptable effects on surface water, ecosystems, land subsidence, or water quality. It is these groundwater systems where disputes over groundwater sustainability typically arise.

The concept of "sustainable development" was brought to the forefront in 1987 by the U.N. World Commission on Environment and Development.[2] The Brundtland Commission, as it came to be known, broadly defined sustainable development as that which meets the needs of the present without compromising the ability of future generations to meet their needs. This straightforward definition doesn't address the many nuances that complicate implementation. While seemingly self-evident, sustainability is among the hardest of goals to achieve. With respect to groundwater resources, sustainability often requires making some very hard choices in the here-and-now, in order to achieve acceptable long-term impacts of groundwater use.

The effects on groundwater depletion, land subsidence, water quality, surface water, and ecosystems are now at a tipping point with many of the world's most critical aquifers. Solving these problems involves much more than simply understanding the facts. Local customs, people's emotional values about the environment, and issues of intergenerational equity all complicate the question of what effects are acceptable. Ideally, establishing groundwater sustainability is a societal decision made through informed public participation. Within this highly complex mix of competing interests, environmental and earth scientists are increasingly challenged to become creative and socially sensitive team players. Simply communicating the facts will no longer get the job done.

Much of the current thinking in addressing groundwater sustainability falls within the broad concept of "groundwater governance," which focuses on promoting responsible collective action.[3] Key factors include:

- Recognizing surface water and groundwater as a single resource
- Active engagement of local stakeholders in the decision-making process
- Pressure from external bodies to achieve suitable and workable solutions
- Public education on groundwater and its importance
- An emphasis on public guardianship and collective responsibility
- Consideration of groundwater within other policy areas, such as agriculture, energy, and land use
- Adequate laws and enforcement
- Fully funded and properly staffed groundwater management agencies
- Characterization of major aquifer systems
- Effective and independent monitoring of groundwater status and trends
- Recognizing the long-term response of groundwater systems
- Accounting for interactions between groundwater and climate
- Community leadership

We emphasize five key issues in closing:

1. *Antiquated groundwater law and policies need to be updated to reflect the dynamic interconnection of surface water and groundwater.* The growing recognition of the need to consider surface water and groundwater as an interrelated system is far from achieved in practice. Separate surface-water and groundwater agencies further undermine holistic resource management. Groundwater considerations also need to be integrated into policies for agriculture, energy, environment, land-use planning, and urban development. Groundwater has an important, but often forgotten, role in these sectors, and decisions within these sectors affect how we use and manage groundwater.

2. *Local solutions are where the action is, but external pressure is usually required to make meaningful progress.* Safeguarding groundwater is a global challenge, but the primary solutions are found at the aquifer, watershed, or local levels. Top-down management usually creates resistance among stakeholders. There's virtually no possibility of getting entrenched groundwater users on board if they aren't actively involved in the decision-making process. At the same time, a carrot-and-stick approach is often needed. Farmers, in particular, abhor regulation. Unlike most stakeholders, it's their livelihood that is directly on the line. An external force is often required to achieve necessary changes and accountability. Yet government by itself cannot solve the problems without public participation and community leadership.

3. *You can't manage what you don't measure.* Quantitative knowledge about groundwater resources is essential for effective and realistic solutions. This is not a simple matter. Data of various types (water-level, water-quality, and geophysical measurements) all require specialized equipment, expertise, and considerable ongoing funding by governmental agencies with sufficient resources and expertise. Data collected over decades are needed to monitor the effects of aquifer development and track long-term trends. While groundwater resources are increasingly promoted as our fallback insurance for drought and the looming specter of climate change, garnering political and financial

support for even the most basic data collection is usually a low priority.

4. *If we wish to manage groundwater resources sustainably, we need to consider the timescale of the consequences.* The response of a groundwater system to pumping—including effects on rivers, lakes, wetlands, and springs—can be many decades. Yet groundwater planning horizons (when they exist) are often only five to twenty years. Some groundwater planning is now looking ahead fifty or even a hundred years, but these longer timeframes typically are associated with deliberate depletion of the resource. For example, a common goal is to maintain groundwater storage at half its current value at the end of fifty years. All too often, we are giving up on making our groundwater resources last beyond a couple generations.

5. *Climate change is simultaneously making groundwater more essential and amplifying threats to this critical resource.* Groundwater becomes even more important for global water and food supply in the face of more frequent and intense droughts and warmer temperatures predicted in climate change scenarios. Climate change will affect groundwater recharge, interactions with surface water, and water quality. The effects are complex. For example, more intense precipitation caused by climate change may make groundwater in the tropics more resilient by increasing recharge.[4] The largest impact, however, is likely to be greater groundwater use. In lieu of this, groundwater basins need to be recharged and conserved during the wet times, in order to prepare for the dry times that inevitably come (with or without climate change).

In Jared Diamond's book *Collapse: How Societies Choose to Fail or Succeed,* he examines how past societies consumed key resources at an unsustainable rate that resulted in environmental suicide. There's even a word for it—"ecocide." Diamond's case studies include Easter Island's complete deforestation, which he cites as one of the most extreme examples of ecocide in the historical record. According to Diamond, the domino effect was rapid and devastating. Lack of large timber brought an end to islanders' ability to construct large seagoing canoes, which destroyed access to their principal meat sources of porpoise, tuna, and pelagic fish. With the trees gone, land birds—another vital food source—completely disappeared, along with the previous abundance of wild fruits such as palm nuts and Malay apples. The absence of trees also meant that the Easter Islanders no longer had fuel with which to keep warm during cold winter nights. Deforestation led to soil erosion, nutrient leaching from the soils that remained, and severely reduced composting materials from leaves, fruit, and twigs. No one knows how long it took (and the exact events are debated), but the consequences of Easter Island's deforestation resulted in starvation, a population crash, and a descent into cannibalism.[5]

Diamond examines other past societies—the Anasazi, the Maya, and Norse Greenlanders—where ecocide directly, or indirectly, resulted in societal collapse. There are also some hopeful stories. The Inuit not only adapted, but actually thrived in the same harsh environment that had destroyed the Norse Greenlanders. Feudal Japan successfully reversed its deforestation, which was reaching the tipping point. A few (very few) others faced their own particular version of ecocide and averted disaster.

Diamond's history addresses past societies whose success or collapse made little or no difference to their neighbors. This is no longer the case. The cumulative effects of today's environmental problems are reaching a level of crisis unprecedented in human history. Globalization now threatens globalized collapse.

In the global context, groundwater is one of our most critical resources. Groundwater has supported an agricultural revolution unlike the world has ever seen. Groundwater is our best insurance against droughts that once resulted in mass starvation. Groundwater is a key pillar of global urbanization, supplying drinking water to the world's megacities. The dry weather flow of rivers and streams, and the health of many lakes and wetlands, depend on groundwater. Yet with all this (and more) at stake, the world's groundwater resources have become critically endangered—both in quantity and quality.

Groundwater, that great hidden resource on which so much depends, is being consumed and contaminated like there's no tomorrow. Yet there is another path to choose. Wisely used, groundwater can continue to help feed the world, alleviate poverty, and provide a source of clean drinking water for much of the world's population, while also maintaining surface-water flows and ecosystems. The choice is ours.

Notes

Chapter 1. Beyond Rain

Epigraph: H. D. Thoreau, *Walden* (New York: Signet Classic, 1960), 189.

1. "Dry Georgia Rallies, and Prays, for Rain," *NBCNews.com,* Nov. 13, 2007.
2. Ibid.
3. S. Inskeep and R. Montagne, "In Drought-Stricken Georgia, a Prayer for Rain," *NPR News, Morning Edition,* Nov. 14, 2007.
4. "Less Water Use in Atlanta Amid Georgia's Water Wars with Alabama, Florida," *Associated Press,* Oct. 19, 2013.
5. N. Hundley, *The Great Thirst: Californians and Water: A History* (Berkeley: University of California Press, 2001), 107–112.
6. Ibid.
7. Ibid.; R. F. Pourade, *Gold in the Sun* (San Diego: Union-Tribune Publishing, 1966); T. W. Patterson, "Hatfield the Rainmaker," *Journal of San Diego History* 16, no. 4 (1970).
8. Z. M. Pike, *Exploratory Travels through the Western Territories of North America Comprising a Voyage from St. Louis, on the Mississippi, to the Source of the River . . . by Order of the Government of the United States* (Denver: W. H. Lawrence, 1889), 230–231.
9. Friends of Homestead National Monument of America, "Homestead Myth: The Rain Follows the Plow," Oct. 16, 2009, available at http://homesteadcongress.blogspot.com/2009/10/homestead-myth-rain-follows-plow.html (accessed May 4, 2016).
10. C. D. Wilber, *The Great Valleys and Prairies of Nebraska and the Northwest* (Omaha: Daily Republican Print, 1881).
11. K. Burns, *The Dust Bowl* (PBS documentary film, 2012).
12. Ibid.
13. W. Ashworth, *Ogallala Blue: Water and Life on the High Plains* (Woodstock, VT: The Countryman, 2006).
14. S. S. Klepfield, "The "Liquid Gold" Rush: Groundwater Irrigation and Law in Nebraska, 1900–93," *Great Plains Quarterly Paper* 738 (1993), available at http://digitalcommons.unl.edu/greatplainsquarterly/738 (accessed May 4, 2016).

15. R. C. Buchanan, R. W. Buddemeier, and B. B. Wilson, *The High Plains Aquifer* (Lawrence: Kansas Geological Survey, 2009).
16. M. J. Jones, *Social Adoption of Groundwater Pumping Technology and the Development of Groundwater Cultures: Governance at the Point of Abstraction,* available at http://www.groundwatergovernance.org/resources/thematic-papers/en/ (accessed May 4, 2016).
17. J. Margat and J. van der Gun, *Groundwater around the World* (Leiden, Neth.: CRC Press/Balkema, 2013).
18. Ibid.
19. Ibid.
20. B. R. Scanlon et al., "Groundwater Depletion and Sustainability of Irrigation in the US High Plains and Central Valley," *Proceedings of the National Academy of Sciences* 109, no. 24 (2012).
21. L. F. Konikow, "Long-Term Groundwater Depletion in the United States," *Groundwater* 53, no. 1 (2015).
22. OECD, *Drying Wells, Rising Stakes: Towards Sustainable Agricultural Groundwater Use* (Paris: OECD Studies on Water, 2015); G. de Marsily, "Green Water, Blue Water, Groundwater: Can We Use Them Conjunctively in a Sustainable Way?" presentation at AQUA2015 42nd IAH Congress, Rome, Sept. 14, 2015.
23. C. Zheng et al., "Can China Cope with Its Water Crisis?—Perspectives from the North China Plain," *Ground Water* 48, no. 3 (2010); J. Yardley, "Beneath Booming Cities, China's Future Is Drying Up," *New York Times,* Sept. 28, 2007; "China Reduces Wheat Irrigation as Farming Depletes Groundwater," *Bloomberg News,* Sept. 15, 2014.
24. J. Barnett et al., "Transfer Project Cannot Meet China's Water Needs," *Nature* 527, no. 7578 (2015).
25. C. Zheng, "Will World's Largest Diversion Solve Deepening Water Crisis of North China Plain?," presentation at AQUA2015 42nd IAH Congress, Rome, Sept. 17, 2015; Y. Jiang, "China's Water Security: Current Status, Emerging Challenges and Future Prospects," *Environmental Science and Policy* 54 (2015).
26. M. Giordano, "Global Groundwater? Issues and Solutions," *Annual Review of Environment and Resources* 34 (2009); D. Seckler, R. Barker, and U. Amarasinghe, "Water Scarcity in the Twenty-First Century," *International Journal of Water Resources Development* 15, nos. 1–2 (1999).
27. B. D. Tapley et al., "GRACE Measurements of Mass Variability in the Earth System," *Science* 305, no. 5683 (2004); F. Barringer, "Groundwater Depletion Is Detected from Space," *New York Times,* May 30, 2011.
28. J. S. Famiglietti and M. Rodell, "Water in Balance," *Science* 340, no. 6138 (2013).

29. W. M. Alley and L. F. Konikow, "Bringing GRACE Down to Earth," *Groundwater* 53, no. 6 (2015).

Chapter 2. India's Silent Revolution

Epigraph: Gandhi is quoted in *Times of India*, Feb. 8, 1959.

1. J. Yardley and G. Harris, "2nd Day of Power Failures Cripples Wide Swath of India," *New York Times,* July 31, 2012.
2. Ibid.
3. "An Elephant, Not a Tiger: A Special Report on India," *Economist,* Dec. 13, 2008.
4. S. Neuman, "India's Power Woes a Classic Story of Supply, Demand," *Minnesota Public Radio News,* July 31, 2012.
5. NGWA, "Groundwater Is Cool," available at https://www.youtube.com/watch?v=ZZ7K50O6z5g (accessed May 4, 2016).
6. J. Margat and J. van der Gun, *Groundwater around the World* (Leiden, Neth.: CRC Press/Balkema, 2013), 126; H. Vaux Jr., "The Economics of Groundwater Resources and the American Experience," in *The Global Importance of Ground Water in the 21st Century,* ed. S. Ragone (Westerville, OH: NGWA Press, 2007), 169.
7. C. Oppenheimer, *Eruptions That Shook the World* (Cambridge, Eng.: Cambridge University Press, 2011), 146–147.
8. T. Shah, M. Giordano, and A. Mukherji, "Political Economy of the Energy-Groundwater Nexus in India: Exploring Issues and Assessing Policy Options," *Hydrogeology Journal* 20, no. 5 (2012).
9. T. Shah, *Taming the Anarchy: Groundwater Governance in South Asia* (New York: Routledge, 2008); K. Schneider, "Scarcity in a Time of Surplus: Free Water and Energy Cause Food Waste and Power Shortage in India," *Circle of Blue,* June 20, 2013.
10. Shah, Giordano, and Mukherji, "Political Economy"; T. Shah, "Climate Change and Groundwater: India's Opportunities for Mitigation and Adaptation," *Environmental Research Letters* 4 (2009).
11. M. R. Llamas and P. Martinez-Santos, "Intensive Groundwater Use: A Silent Revolution That Cannot Be Ignored," *Water Science and Technology* 51, no. 8 (2005).
12. *Summary: Comparative Study on Impact of Flat Rate Tariff and Metered Tariff in the Agricultural Sector* (New Delhi: Rural Electrification Corporation, 1985).
13. M. J. Jones, *Social Adoption of Groundwater Pumping Technology and the Development of Groundwater Cultures; Governance at the Point of Abstraction,* available at http://www.groundwatergovernance.org/resources

/thematic-papers/en/ (accessed May 5, 2016); Shah, Giordano, and Mukherji, "Political Economy."

14. Shah, Giordano, and Mukherji, "Political Economy."
15. *Deep Wells and Prudence: Towards Pragmatic Action for Addressing Groundwater Overexploitation in India* (Washington, DC: World Bank, 2010).
16. A. Sarkar, "Socio-Economic Implications of Depleting Groundwater Resource in Punjab: A Comparative Analysis of Different Irrigation Systems," *Economics and Political Weekly,* Feb. 12, 2011.
17. Quotes from Shah, *Taming the Anarchy.*
18. Shah, Giordano, and Mukherji, "Political Economy."
19. M. R. Llamas and P. Martinez-Santos, "Intensive Groundwater Use: Silent Revolution and Potential Source of Social Conflicts," *Journal of Water Resources Planning and Management* 131, no. 5 (2005).
20. "India Faces Severe Water Crisis in 20 Years: World Bank," *TerraDaily,* Oct. 5, 2005, available at http://www.terradaily.com/news/water-earth-05ze.html (accessed May 5, 2016).
21. T. Shah, "Mobilising Social Energy against Environmental Challenge: Understanding the Groundwater Recharge Movement in Western India," *Natural Resources Forum* 24, no. 3 (2000); F. van Steenbergen, "Promoting Local Management in Groundwater," *Hydrogeology Journal* 14, no. 3 (2006).
22. Shah, "Mobilising Social Energy."
23. Ibid.
24. S. Mudrakartha, *Adaptive Approaches to Groundwater Governance: Lessons from the Saurashtra Recharging Movement* (Anand [Gujarat], India: Institute of Rural Management Anand, 2008).
25. T. Shah et al., "Groundwater Governance through Electricity Supply Management: Assessing an Innovative Intervention in Gujarat, Western India," *Agricultural Water Management* 95, no. 11 (2008).
26. Ibid.; *12th Five Year Plan* (New Delhi: Planning Commission, Government of India, 2012).
27. R. K. Singh, "India Turns to Irrigation Pumps to Ease Crippling Power Debt," *Bloomberg Business,* Jan. 21, 2016; S. M. Khair, S. Mushtaq, and K. Reardon-Smith, "Groundwater Governance in a Water-Starved Country: Public Policy, Farmer's Perceptions, and Drivers of Tubewell Adoption in Balochistan, Pakistan," *Groundwater* 53, no. 4 (2015); C. A. Scott, "Electricity for Groundwater Use: Constraints and Opportunities for Adaptive Response to Climate Change," *Environmental Research Letters* 8, no. 3 (2013).

Chapter 3. Arizona

Epigraph: J. Steinbeck, *East of Eden* (New York: Viking, 1952).

1. M. C. Carpenter, "South-Central Arizona," in *Land Subsidence in the United States,* ed. D. Galloway, D. R. Jones, and S. E. Ingebritsen (Reston, VA: U.S. Geological Survey, 1999), 65–78.
2. Ibid.
3. J. L. August and G. Gammage, "Shaped by Water: An Arizona Historical Perspective," in *Arizona Water Policy: Management Innovations in an Urbanizing, Arid Region,* ed. B. G. Colby and K. L. Jacobs (Washington, DC: RFF Press, 2007), 10–25.
4. N. Hundley, *The Great Thirst: Californians and Water; A History* (Berkeley: University of California Press, 2001).
5. Ibid.
6. Ibid.
7. J. Kightlinger, "Water in 2050: The Murky Crystal Ball," *Water Resources IMPACT* 16, no. 1 (2014).
8. Hundley, *Great Thirst.*
9. August and Gammage, "Shaped by Water."
10. N. Hundley, "The West against Itself," in *New Courses for the Colorado River: Major Issues for the Next Century,* ed. G. D. Weatherford and F. L. Brown (Albuquerque: University of New Mexico Press, 1986), 25.
11. August and Gammage, "Shaped by Water."
12. M. Reisner, *Cadillac Desert: The American West and Its Disappearing Water* (New York: Penguin, 1993), 260–262.
13. Hundley, *Great Thirst;* ibid.
14. Personal interview with Kathleen Ferris, Apr. 2, 2014; B. Leverton, "Interview with Kathy Ferris, CAP Oral Histories, May 23, 2005," available at http://www.cap-az.com/documents/about/oral-histories/Interview_with_Kathy_Ferris.pdf (accessed May 4, 2016).
15. L. W. Staudenmaier, "Between a Rock and a Dry Place: The Rural Water Supply Challenge for Arizona," *Arizona Law Review* 49, no. 2 (2007).
16. S. B. Megdal, "Arizona Groundwater Management," *Water Report,* Oct. 15, 2012.
17. Ibid.
18. "Central Arizona Project," available at http://cap-az.com (accessed May 9, 2016).
19. T. James et al., *The Economic Impact of the Central Arizona Project to the State of Arizona* (Tempe: Arizona State University, 2014).
20. J. Gelt et al., *Water in the Tucson Area: Seeking Sustainability* (Tucson: Arizona Water Resources Research Center, 1999).
21. Ibid.

22. S. B. Megdal and A. Forrest, "How a Drought-Resilient Water Delivery System Rose Out of the Desert: The Case of Tucson Water," *Journal of the American Water Works Association* 107, no. 9 (2015).
23. C. R. Schwalm et al., "Reduction in Carbon Uptake during Turn of the Century Drought in Western North America," *Nature Geosciences* 5 (2012).
24. *Arizona Water Banking Authority Annual Report, 2013* (Phoenix: Arizona Water Banking Authority, 2014); S. Eden, M. Ryder, and M. A. Capehart, "Closing the Water Demand-Supply Gap in Arizona," *Arroyo* (Tucson: University of Arizona Water Resources Research Center, 2015).
25. Megdal, "Arizona Groundwater Management"; the Mulroy quotation is in S. McClurg, "Solving the Colorado River Basin's Math Problem: Adapting to Change," *Western Water* (Nov.–Dec. 2011).
26. Megdal, "Arizona Groundwater Management."
27. D. R. Smith and B. G. Colby, "Tribal Water Claims and Settlements within Regional Water Management," in *Arizona Water Policy: Management Innovations in an Urbanizing, Arid Region,* ed. B. G. Colby and K. L. Jacobs (Washington, DC: RFF Press, 2007), 204–218.
28. Megdal, "Arizona Groundwater Management"; C. A. Avery et al., "Good Intentions, Unintended Consequences: The Central Arizona Groundwater Replenishment District," *Arizona Law Review* 49, no. 2 (2007); K. Ferris, "New Plan Doesn't Fix Groundwater Loophole," *Arizona Republic—AZCentral.com,* Dec. 4, 2014.
29. S. Eden, M. Efrein, and L. Radonic, "What Is the Value of Water? A Complex Question" (Tucson: University of Arizona Water Resources Research Center, 2014).
30. Eden, Ryder, and Capehart, "Closing the Water Demand-Supply Gap."

Chapter 4. The World's Poorest People

Epigraph: B. Richter, "Chasing Water in a Rapidly Changing World," Presentation at University of Arizona Water Resources Research Center, Mar. 5, 2015.

1. "Data, Sub-Saharan Africa," available at http://data.worldbank.org/region/sub-saharan-africa (accessed May 8, 2016).
2. M. Fleshman, "Laying Africa's Roads to Prosperity," *Africa Renewal Online* (Jan. 2009).
3. "Data, Sub-Saharan Africa."
4. UNICEF and WHO, *Progress on Drinking Water and Sanitation: 2015 Update and MDG Assessment* (Geneva: World Health Organization, 2015).
5. "Millions Lack Safe Water," available at http://water.org/water-crisis/water-facts/water/ (accessed May 8, 2016).

6. L Liu et al., "Global, Regional, and National Causes of Child Mortality in 2000–13, with Projections to Inform Post-2015 Priorities: An Updated Systematic Analysis," *Lancet* 385, no. 9966 (2015).
7. UNICEF and WHO, *Progress on Drinking Water and Sanitation: 2015;* United Nations, *The Millennium Development Goals Report, 2015* (New York: United Nations, 2015).
8. Ibid.
9. WaterAid, *Water: At What Cost? The State of the World's Water, 2016 (London: WaterAid, 2016).*
10. D. Hulme, *The Millennium Development Goals (MDGs): A Short History of the World's Biggest Promise* (Manchester, Eng.: University of Manchester, 2009).
11. Ibid.
12. Ibid.; L. Emmerij, R. Jolly, and T. G. Weiss, *Ahead of the Curve? UN Ideas and Global Challenges* (Bloomington, IN: Indiana University Press, 2001).
13. K. A. Annan, *We the Peoples: The Role of the United Nations in the 21st Century* (New York: United Nations Department of Public Information, 2000).
14. United Nations, *Millennium Development Goals Report, 2015.*
15. Rural Water Supply Network (RWSN), "Human Right to Water: What Does It Mean in Practice?" available at http://rural-water-supply.net/en/resources/details/503 (accessed May 4, 2016).
16. Ibid.
17. UNICEF and WHO Joint Monitoring Program, *Progress on Drinking Water and Sanitation: 2012 Update* (Geneva: World Health Organization, 2012).
18. K. Harmon, "Improved But Not Always Safe: Despite Global Efforts, More Than 1 Billion People Likely at Risk for Lack of Clean Water," *Scientific American,* May 21, 2012; The 2015 UN/WHO estimate is that 663 million people lack access to 'improved' water sources.
19. P. Iyer, J. Davis, and E. Yavuz, *Rural Water Supply, Sanitation, and Hygiene: A Review of 25 Years of World Bank Lending (1978–2003)—Summary Report* (New York: World Bank, 2006); RWSN Executive Steering Committee, "Myths of the Rural Water Supply Sector," Rural Water Supply Network (RWSN), available at http://www.rural-water-supply.net/en/resources/details/226 (accessed May 4, 2016).
20. RWSN Executive Steering Committee, "Myths of the Rural Water Supply Sector"; K. Purvis, "How Do You Solve a Problem Like a Broken Water Pump?," *Guardian,* Mar. 22, 2016.
21. "Sustainable Development Goals," available at https://sustainabledevelopment.un.org/sdgs (accessed May 4, 2016).

22. "Rural Water Supply Network (RWSN)," available at http://www.rural-water-supply.net/en/; K. Danert, "Experiences and Ideas from RWSN's Sustainable Groundwater Community 2013," available at http://www.rural-water-supply.net/en/resources/details/517 (accessed May 4, 2016); K. Danert et al., *Code of Practice for Cost Effective Boreholes* (St. Gallen, Switz.: Rural Water Supply Network, 2010).
23. R. H. Holm, "Recent History Provides Sustainable African Water Quality Project Insight," *Ground Water* 50, no. 5 (2012).
24. The Ryan Hreljac quotations are from M. Price, "Ryan Hreljac, Founder of Ryan's Well Foundation," *Water Well Journal* 68, no. 11 (2014).
25. S. J. Schneider, "Water Supply Well Guidelines for Use in Developing Countries," available at http://seidc.com/pdf/Hydrophilanthropy_Well_Guidelines.pdf (accessed May 4, 2016); "Ann Campana Judge Foundation," available at http://acjfoundation.org/ (accessed May 4, 2016).
26. Personal interview with Paul Polak, Oct. 24, 2013; P. Polak, *Out of Poverty* (San Francisco: Berrett-Koehler, 2008); and P. Polak and M. Warwick, *The Business Solution to Poverty: Designing Products and Services for Three Billion New Customers* (San Francisco: Berrett-Koehler, 2013).
27. P. Polak, "From Groundwater to Wealth for One-Acre Farmers," in *The Global Importance of Ground Water in the 21st Century,* ed. S. Ragone (Westerville, OH: NGWA Press, 2006), 205–218.

Chapter 5. Not All Aquifers Are Created Equal

Epigraph: Miller is quoted in National Research Council, *Ground Water Vulnerability Assessment* (Washington, DC: National Academy Press, 1993), 63.

1. "National Water Precious Resource Map," *National Geographic Magazine,* Nov. 1993.
2. W. M. Alley, "Tracking U.S. Groundwater: Reserves for the Future?," *Environment* 48, no. 3 (2006).
3. The term "artesian aquifer" is sometimes used as a synonym for confined aquifer, whether or not wells tapping the aquifer are flowing wells. We use the term "artesian well" to refer specifically to a flowing well.
4. N. Ball, "The Other Great Nineteenth-Century Tower of Paris," *Parisian Fields,* Oct. 28, 2012, available at https://parisianfields.wordpress.com/2012/10/28/the-other-great-tower-of-paris/ (accessed May 4, 2016); C. Contoux, "How Basin Model Results Enable the Study of Multi-Layer Response to Pumping: The Paris Basin, France," *Hydrogeology Journal* 21, no. 3 (2013).
5. R. E. Mace, "So Secret, Occult, and Concealed: The Story of Groundwater in Texas," Presentation at the 2015 NGWA Groundwater Summit, San Antonio, Texas, Mar. 17, 2015.

6. H. W Bentley et al., "Chlorine 36 Dating of Very Old Groundwater 1. The Great Artesian Basin, Australia," *Water Resources Research* 22, no. 13 (1986).
7. M. Perry, "Ancient Water Source Vital for Australia," *Reuters,* Dec. 23, 2008; J. Margat and J. van der Gun, *Groundwater around the World* (Leiden, Neth.: CRC Press/Balkema, 2013), 89.
8. O. Powell et al., "Oases to Oblivion: The Rapid Demise of Springs in the South-Eastern Great Artesian Basin, Australia," *Groundwater* 53, no. 1 (2015); Perry, "Ancient Water Source."
9. Government of South Australia, "Natural Resources SA Arid Lands," available at http://www.naturalresources.sa.gov.au/aridlands/water /managing-water-resources/ground-water (accessed May 4, 2016).
10. A. M. MacDonald et al., "Quantitative Maps of Groundwater Resources in Africa," *Environmental Research Letters* 7, no. 2 (2012); J. Chilton, "Groundwater in Africa: Research Article Generates Intense Media Interest," *International Association of Hydrogeologists Newsletter* (Aug. 2012); A. Laing, "African Continent 'Sitting on Vast Reservoir,'" *Daily Telegraph,* Apr. 20, 2012; C. Wickham, "Africa Sitting on Sea of Groundwater Reserves," *Reuters,* Apr. 20, 2012.
11. D. E. Walling, "Hydrology and Rivers," in *The Physical Geography of Africa,* ed. W. Adams, A. Goudie, and A. Orme (Oxford, Eng.: Oxford University Press, 1996).
12. MacDonald et al., "Quantitative Maps of Groundwater Resources in Africa"; A. M. MacDonald et al., "What Impact Will Climate Change Have on Rural Groundwater Supplies in Africa?" *Hydrological Sciences Journal* 54, no. 4 (2009); E. Braune and Y. Xu, "The Role of Ground Water in Sub-Saharan Africa," *Ground Water* 48, no. 2 (2010).
13. M. A. Maupin and N. L. Barber, *Estimated Withdrawals from Principal Aquifers in the United States, 2000* (Reston, VA: U.S. Geological Survey, 2005).
14. M. Giordano, "Agricultural Groundwater Use and Rural Livelihoods in Sub-Saharan Africa: A First-Cut Assessment," *Hydrogeology Journal* 14, no. 3 (2006); K. G. Villholth, "Groundwater Irrigation for Smallholders in Sub-Saharan Africa—A Synthesis of Current Knowledge to Guide Sustainable Outcomes," *Water International* 38, no. 4 (2013).
15. S. Foster and D. P. Loucks, ed. *Non-Renewable Groundwater Resources: A Guidebook on Socially Sustainable Management for Water-Policy Makers,* Series on Groundwater, no. 10 (Paris: UNESCO-IHP-VI, 2006).
16. C. I. Voss and S. M. Soliman, "The Transboundary Non-Renewable Nubian Aquifer System of Chad, Egypt, Libya and Sudan: Classical Groundwater Questions and Parsimonious Hydrogeologic Analysis and Modeling," *Hydrogeology Journal* 22, no. 2 (2014).

17. "Libya's Water Supply: Plumbing the Sahara," *Economist,* Mar. 11, 2011.
18. J. Watkins, "Libya's Thirst for 'Fossil Water,'" *BBC World Service,* Mar. 18, 2006.
19. S. A. Topol, "Libya's Qaddafi Taps 'Fossil Water' to Irrigate Desert Farms," *Christian Science Monitor,* Aug. 23, 2010.
20. S. Postel, *Last Oasis: Facing Water Scarcity* (New York: W.W. Norton, 1992); O. K. M. Ouda, "Impacts of Agricultural Policy on Irrigation Water Demand: A Case Study of Saudi Arabia," *International Journal of Water Resources Development* 30, no. 2 (2014); Margat and van der Gun, *Groundwater around the World,* 132.
21. R. Glennon, *Unquenchable: America's Water Crisis and What to Do about It* (Washington, DC: Island Press, 2009).
22. G. Eckhardt, "Ron Pucek's Living Waters Catfish Farm," available at http://www.edwardsaquifer.net/pucek.html (accessed May 4, 2016).
23. Ibid.
24. T. H. Votteler, "Raiders of the Lost Aquifer? or, The Beginning of the End to Fifty Years of Conflict over the Texas Edwards Aquifer," *Tulane Environmental Law Journal* 15, no. 2 (2002).
25. G. Eckhardt, "The Edwards Aquifer Website," available at http://www.edwardsaquifer.net (accessed May 4, 2016).
26. N. Jackson, "Mermaid Theatre," *Popular Mechanics* (June 1952).
27. G. Longley, "The Edwards Aquifer: Earth's Most Diverse Groundwater Ecosystem?," *International Journal of Speleology* 11, no. 1 (1981).
28. Votteler, "Raiders of the Lost Aquifer?"
29. W. Back and R. A. Freeze, ed., *Chemical Hydrogeology* (Stroudsburg, PA: Hutchinson Ross, 1983), 90.
30. P. M. Barlow, *Ground Water in Freshwater-Saltwater Environments of the Atlantic Coast* (Reston, VA: U.S. Geological Survey, 2003).
31. Hilton Head Public Service District, "Saltwater Intrusion and Hilton Head's Response," available at http://www.hhpsd.com/saltwater-intrusion/ (accessed May 8, 2016).
32. "Overdraft and Saltwater Intrusion Strain the Floridan Aquifer," *Circle of Blue,* Nov. 5, 2010.
33. I. White and T. Falkland, "Saltwater Intrusion in Coastal Regions of North America," *Hydrogeology Journal* 18, no. 1 (2010); R. T. Bailey and J. W. Jenson, "Effects of Marine Overwash for Atoll Aquifers: Environmental and Human Factors," *Groundwater* 52, no. 5 (2014).
34. V. E. A. Post et al., "Offshore Fresh Groundwater Reserves as a Global Phenomenon," *Nature* 504, no. 7478 (2013).

35. J. S. Stanton, "Importance, Distribution, and Character of the Nation's Brackish Groundwater Resources," Presentation at the 2016 NGWA Groundwater Summit, Denver, Colorado, Apr. 24–27, 2016; LBG-Guyton Associates, *Brackish Groundwater Manual for Texas Regional Water Planning Groups* (Austin: Texas Water Development Board, 2003).

Chapter 6. Who Owns Groundwater?

Epigraph: M. N. Goodman, "Current Groundwater Law in Arizona," *Arizona State Law Journal* 2–3 (1978).

1. N. Hundley, *The Great Thirst: Californians and Water, a History* (Berkeley: University of California Press, 2001), 528.
2. *Frazier v. Brown,* 12 Ohio St. 294 (1861).
3. R. Kaiser, "Who Owns the Water? A Primer on Texas Groundwater and Spring Flow," *Texas Parks and Wildlife* (July 2005), available at http://www.tpwmagazine.com/archive/2005/jul/ed_2 (accessed May 4, 2016).
4. N. Satija, "Groundwater Wars Brewing in Austin's Suburbs," *Texas Tribune,* Jan. 23, 2015.
5. N. Satija, "Hays County Groundwater Bill Heads to Governor's Desk," *Texas Tribune,* May 31, 2015.
6. D. Cruse, "Landmark Texas Water Rights Case May Lead to Future Takings Claims or Legislative Fixes: *Edwards Aquifer v. Day,*" *Supreme Court of Texas Blog,* Feb. 24, 2012; C. McDonald, "Who Owns Groundwater in the Aquifer?" *San Antonio Express-News,* Feb. 18, 2010.
7. Cruse, "Landmark Texas Water Rights Case."
8. N. Satija, "Texas Groundwater Districts Face Bevy of Challenges," *Texas Tribune,* Aug. 29, 2013; J. Malewitz, "State Supreme Court Punts on Major Water Case," *Texas Tribune,* May 1, 2015.
9. W. Ashworth, *Ogallala Blue: Water and Life on the High Plains* (Woodstock, VT: The Countryman, 2006).
10. Ibid.; S. Berfield, "There Will Be Water," *Bloomberg Businessweek,* June 11, 2008.
11. Berfield, "There Will Be Water."
12. L. Woellert, "Pickens Water Plan Poised to Gain Bond, Condemnation Authority," *Bloomberg,* Nov. 6, 2007.
13. Berfield, "There Will Be Water."
14. Hundley, *The Great Thirst.*
15. R. D. Benson, "A Few Ironies of Western Water Law," *Wyoming Law Review* 6, no. 2 (2006).
16. J. Fleck, "Ruling Protects Senior Water Rights," *Albuquerque Journal,* July 26, 2013.

17. J. Horwath and J. Peters, "Well, Well—What the State Supreme Court's New Water-rights Ruling Means for New Mexicans," *Santa Fe Reporter,* July 30, 2013.
18. S. E. Reynolds, *Twenty-Fifth Biennial Report of the State Engineer of New Mexico* (Albuquerque: The Valliant Company, 1962), 91; M. S. Johnson, "Deep Groundwater Administration in New Mexico," Presentation at the NGWA Conference on Characterization of Deep Groundwater, Denver, Colorado, May 8, 2014.
19. Johnson, "Deep Groundwater Administration in New Mexico."
20. M. D. Dettinger et al., "Atmospheric Rivers, Floods and the Water Resources of California," *Water* 3, no. 2 (2011); W. C. Mendenhall, *Ground Waters and Irrigation Enterprises in the Foothill Belt, Southern California* (Washington, DC: U.S. Geological Survey, 1908).
21. Hundley, *The Great Thirst.*
22. V. K. Grabert and T. N. Narasimhan, "California's Evolution toward Integrated Regional Water Management: A Long View," *Hydrogeology Journal* 14, no. 3 (2006).
23. R. L. Nelson, "Assessing Local Planning to Control Groundwater Depletion: California as a Microcosm of Global Issues," *Water Resources Research* 48, no. 1 (2012).
24. S. Hockensmith, "Why the State's Water Woes Could Be Just Beginning," *UC Berkeley NewsCenter,* Jan. 21, 2014.
25. I. Lovett, "Parched, California Cuts off Tap to Agencies," *New York Times,* Jan. 31, 2015; "California Farmers Told Not to Expect U.S. Water," *New York Times,* Feb. 22, 2015.
26. J. Gillis and M. Richtel, "Beneath California Crops, Groundwater Crisis Grows," *New York Times,* Apr. 5, 2015.
27. Governor of the State of California, *California Water Action Plan* (Sacramento: Governor's Office, 2004); "Governor's Budget Summary—2014–15," available at http://www.ebudget.ca.gov/2014–15/pdf/BudgetSummary/NaturalResources.pdf (accessed Mar. 1, 2015).
28. Editorial Board, "California Needs Overdraft Protection for Its Dwindling Groundwater Supplies," *Sacramento Bee,* Apr. 13, 2014.
29. J. Margat and J. van der Gun, *Groundwater around the World* (Leiden, Neth.: CRC Press/Balkema, 2013), 149.
30. M. R. Llamas et al., "Groundwater in Spain: Increasing Role, Evolution, Present and Future," *Environmental Earth Sciences* 73, no. 6 (2015).
31. M. R. Llamas and E. Custodio, "Intensive Use of Groundwater: A New Situation Which Demands Proactive Action," in *Intensive Use of Ground-*

water Challenges and Opportunities, ed. M. R. Llamos and E. Custodio (Leiden, Neth.: CRC Press/Balkema, 2003), 13–31.

32. E. Esteban and J. Albiac, "The Problem of Sustainable Groundwater Management: The Case of La Mancha Aquifers, Spain," *Hydrogeology Journal* 20, no. 5 (2012).
33. M. R. Llamas, "Conflicts between Wetland Conservation and Groundwater Exploitation: Two Case Histories in Spain," *Environmental Geology and Water Sciences* 11, no. 3 (1988); L. Moreno et al., "The 2009 Smouldering Peat Fire in Las Tablas de Daimiel National Park (Spain)," *Fire Technology* 47, no. 2 (2011).
34. Esteban and Albiac, "Problem of Sustainable Groundwater Management."
35. Ibid.; Moreno et al., "2009 Smouldering Peat Fire."
36. Esteban and Albiac, "Problem of Sustainable Groundwater Management."
37. N. Harrington and P. Cook, *Groundwater in Australia* (Adelaide: National Centre for Groundwater Research and Training, 2014); J. Peel and J. Choy, *Water Governance and Climate Change: Drought in California as a Lens of Our Climate Future* (Stanford, CA: Water in the West, 2014).
38. B. Richter, *Chasing Water* (Washington, DC: Island Press, 2014); C. Fishman, *The Big Thirst* (New York: Free Press, 2011), 146.
39. "The Big Dry," *Economist,* Apr. 26, 2007; T. L. Friedman, "The Aussie 'Big Dry,'" *New York Times,* May 4, 2007.
40. D. Connell, "Irrigation, Water Markets and Sustainability in Australia's Murray-Darling Basin," *Agriculture and Agricultural Science Procedia* 4 (2015); National Water Commission, *Intergovernmental Agreement on a National Water Initiative* (Canberra: Commonwealth of Australia, 2004).
41. Natural Resources Management Standing Committee, *Groundwater Trading* (Canberra: Commonwealth of Australia, 2002).
42. Richter, *Chasing Water;* A. Aghakouchak et al., "Australia's Drought: Lessons for California," *Science* 343, no. 6178 (2014).
43. "Howard Pushes for Control of Murray-Darling," *Australian,* Jan. 25, 2006.
44. J. Horne, "The 2012 Murray-Darling Basin Plan—Issues to Watch," *International Journal of Water Resources Development* 30, no. 1 (2014).

Chapter 7. Streamflow Depletion

Epigraph: Lewis, "The Pecos River in New Mexico: Lessons for the Rio Grande (or Not!)," Presentation at NGWA Conference on Hydrology and Water Scarcity in the Rio Grande Basin, Albuquerque, NM, Feb. 25–26, 2014.

1. "Swinomish Indian Tribal Community," available at http://www.swinomish-nsn.gov/ (accessed May 8, 2016); K. Martin, "Skagit's Water Rights Showdown," *Skagit Valley Herald,* Apr. 23, 2012.

2. D. L. Timmons, "Water Rights: Washington Supreme Court Refuses to Allow New Diversions to Reduce In-Stream Flows, Puts Developers, Homeowners in Bind," *Marten Law Newsletter,* Oct. 27, 2013.
3. R. Lerman, "Ecology Seeks to Ease Tensions at Water Meeting," *Skagit Valley Herald,* Dec. 4, 2013; "Skagit River Basin—Water Management Rule," available at http://www.ecy.wa.gov/programs/wr/instream-flows/skagitbasin.html (accessed May 8, 2016).
4. S. N. Davis and R. J. M. DeWeist, *Hydrogeology* (New York: John Wiley & Sons, 1966).
5. A. N. Whitehead, *Science and the Modern World* (New York: Free Press, 1967), 6; D. Deming, "Born to Trouble: Bernard Palissy and the Hydrologic Cycle," *Ground Water* 43, no. 6 (2005).
6. Quoted in Deming, "Born to Trouble."
7. D. Deming, "Pierre Perrault, the Hydrologic Cycle and the Scientific Revolution," *Groundwater* 52, no. 1 (2014).
8. Davis and DeWeist, *Hydrogeology.*
9. L. F. Konikow and S. A. Leake, "Depletion and Capture: Revisiting 'The Source of Water Derived from Wells,'" *Groundwater* 52, no. 1 (2014).
10. Colorado Foundation for Water Education, "A Decade of Colorado Supreme Court Decisions: 1996–2006," *Headwaters* (Fall 2007).
11. R. Cantwell, ed., *Citizen's Guide to Colorado's Interstate Compacts* (Denver: Colorado Foundation for Water Education, 2010).
12. W. M. Alley, "Tracking U.S. Groundwater: Reserves for the Future?," *Environment* 48, no. 3 (2006).
13. Colorado Division of Water Resources, "Republican River Compact Details," available at http://water.state.co.us/SurfaceWater/Compacts/RepublicanRiver/Pages/RepublicanRiverCompact.aspx (accessed May 5, 2016).
14. E. McIntyre, "A River in Debt," *Headwaters* (Summer 2006).
15. Cantwell, *Citizen's Guide.*
16. "Bonny Lake Drained, Thousands of Fish Die," *9NEWS.com,* Mar. 2, 2014.
17. J. D. Aiken, "The Western Common Law of Tributary Groundwater: Implications for Nebraska," *Nebraska Law Review* 83, no. 2 (2004).
18. E. Schlager et al., "The Costs of Compliance with Interstate Agreements: Lessons from Water Compacts in the Western United States," *Publius: The Journal of Federalism* 42, no. 3 (2012).
19. Kansas Department of Agriculture, *Republican River Compact Enforcement Fact Sheet* (Topeka: Kansas Department of Agriculture, 2008).
20. B. Walton, "In Sign of the Times, A Water Pipeline in Nebraska Taps the Ogallala to Serve Thirsty Kansas," *Circle of Blue,* Jan. 19, 2014.

21. J. P. Jacobs, "Supreme Court Rules Neb. Must Pay Kan. in Interstate River Battle," *E&E Publishing, LLC,* Feb. 24, 2015.
22. A. Bleed and C. H. Babbitt, *Nebraska's Natural Resources Districts* (Lincoln: Robert B. Daugherty Institute, University of Nebraska, 2015).
23. J. Smith, "Aquifers in Free Fall," *Headwaters* (Summer 2013).
24. P. A. Jones, "South Platte Well Crisis, 2002–2010," *Water Report,* Aug. 15, 2010.
25. R. A. Young and J. D. Bredehoeft, "Digital Computer Simulation for Solving Management Problems of Conjunctive Ground and Surface Water Systems," *Water Resources Research* 8, no. 3 (1972).
26. L. Strawn, "The Last GASP: The Conflict over Management of Replacement Water in the South Platte River Basin," *University of Colorado Law Review* 75, no. 2 (2004).
27. Colorado State University (CSU), "Colorado Water Special Edition: South Platte Groundwater," *Newsletter of the CSU Water Center* 31, no. 1 (2014).
28. Ibid.; L. Ozzello, "South Platte Well Owners in Crisis," *Headwaters* (Summer 2006).
29. Ozzello, "South Platte Well Owners."
30. Cech quoted in Ozzello, "South Platte Well Owners."
31. Jones, "South Platte Well Crisis"; Colorado State University, "Colorado Water Special Edition"; Ozzello, "South Platte Well Owners."
32. Jones, "South Platte Well Crisis."
33. D. Martinez, "Colorado Water Institute Studying Groundwater Issues on South Platte," *Sterling Journal-Advocate,* Jan. 15, 2013.
34. J. Bredehoeft, "Hydrologic Trade-offs in Conjunctive Use Management," *Ground Water* 49, no. 4 (2011).
35. B. Walton, "Wisconsin Groundwater Dispute Is a Warning Signal for the Eastern United States," *Circle of Blue,* Oct. 26, 2015.

Chapter 8. Water for Nature

Epigraph: The Nature Conservancy, ad at San Diego International Airport, Dec. 2014.

1. Theodore Roosevelt Association, "The Conservationist," available at www.theodoreroosevelt.org (accessed May 5, 2016).
2. Ibid.
3. USDA Forest Service, *Water and the Forest Service* (Washington, DC: USDA Forest Service, 2000).
4. *Federal Register* 79, no. 87 (May 6, 2014): 25815–25824.
5. M. Griswold, "Groundwater Experts, Water Law Experts, and Conservation Groups Tell the Forest Service to Do More to Protect Groundwater," *Natural Resources Defense Council Staff Blog,* Oct. 14, 2014.

6. Western Governors' Association, "Groundwater Directive Comments," letter from B. Sandoval and J. Kitzhaber to Elizabeth Berger, Oct. 2, 2014.
7. S. Glasser et al., *Technical Guide to Managing Ground Water Resources* (Washington, DC: USDA Forest Service, 2007), 12.
8. U.S. Supreme Court, *Cappaert v. United States,* 426 U.S. 128 (1976).
9. P. A. Byorth, "Conflict to Compact: Federal Reserved Water Rights, Instream Flows, and Native Fish Conservation on National Forests in Montana," *Public Land and Resources Review* 30 (2009).
10. N. Funke et al., "Redressing Inequality, South Africa's New Water Policy," *Environment* 49, no. 3 (2007).
11. Ibid.
12. Quoted in B. van Koppen and B. Schreiner, "Moving beyond Integrated Water Resource Management: Developmental Water Management in South Africa," *International Journal of Water Resources Development* 30, no. 3 (2014).
13. Byorth, "Conflict to Compact"; S. Postel and B. Richter, *Rivers for Life: Managing Water for People and Nature* (Washington, DC: Island Press, 2003).
14. H. M. MacKay et al., "Implementing the South African Water Policy: Holding the Vision While Exploring an Uncharted Mountain," *Water S.A.* 29, no. 40 (2003); J. Levy and Y. Xu, "Review: Groundwater Management and Groundwater/Surface-Water Interaction in the Context of South African Water Policy," *Hydrogeology Journal* 20, no. 2 (2012).
15. United Nations Development Programme (UNDP), *Beyond Scarcity: Power, Poverty, and the Global Water Crisis* (New York: UNDP, 2006).
16. MacKay et al., "Implementing the South African Water Policy"; E. van Wyk et al., "The Ecological Reserve: Towards a Common Understanding for River Management in South Africa," *Water S.A.* 32, no. 3 (2006); P. Seward, "Challenges Facing Environmentally Sustainable Ground Water Use in South Africa," *Ground Water* 48, no. 2 (2010).
17. Funke et al., "Redressing Inequality."
18. Van Koppen and Schreiner, "Moving beyond Integrated Water Resource Management."
19. United Nations Development Programme, *Beyond Scarcity;* K. Schneider, "Drought Pushes South Africa To Water, Energy, and Food Reckoning," *Circle of Blue,* Jan. 19, 2016.
20. A. Springer, "Importance of Springs to Humans," Feb. 20, 2015, available at http://voices.nationalgeographic.com/2015/02/20/springs-the-canary-in-a-coal-mine-for-groundwater/ (accessed May 8, 2016); B. Kløve et al., "Groundwater Dependent Ecosystems. Part 1: Hydroecological Status and Trends," *Environmental Science & Policy* 14, no. 7 (2011).

21. G. Power, R. S. Brown, and J. G. Imhof, "Groundwater and Fish—Insights from Northern North America," *Hydrological Processes* 13, no. 3 (1999).
22. S. Kemper, "Perilous Journeys," *National Wildlife,* Oct. 1, 2008, available at http://www.nwf.org/News-and-Magazines/National-Wildlife/Birds/Archives/2008/Perilous-Journeys.aspx (accessed May 5, 2016).
23. Ibid.
24. R. Glennon, *Water Follies* (Washington, DC: Island Press, 2002).
25. Ibid.; R. H. Webb, S. A. Leake, and R. M. Turner, *The Ribbon of Green* (Tucson: University of Arizona Press, 2007).
26. H. E. Richter et al., "Development of a Shared Vision for Groundwater Management to Protect and Sustain Baseflows of the Upper San Pedro River, Arizona, USA," *Water* 6, no. 8 (2014).
27. Glennon, *Water Follies;* G. Saliba and K. L. Jacobs, "Saving the San Pedro River," *Environment* 50, no. 6 (2008).
28. Glennon, *Water Follies,* 57.
29. Saliba and Jacobs, "Saving the San Pedro River."
30. The Holly Richter quotations are from a personal interview on Apr. 3, 2014.
31. Saliba and Jacobs, "Saving the San Pedro River"; D. R. Pool and J. E. Dickinson, *Ground-water Flow Model of the Sierra Vista Subwatershed and Sonoran Portions of the Upper San Pedro Basin, Southeastern Arizona, United States, and Northern Sonora, Mexico* (Reston, VA: U.S. Geological Survey, 2007).
32. S. A. Leake, D. R. Pool, and J. M. Leenhouts, *Simulated Effects of Ground-Water Withdrawals and Artificial Recharge on Discharge to Streams, Springs, and Riparian Vegetation in the Sierra Vista Subwatershed of the Upper San Pedro Basin, Southeastern Arizona* (Reston, VA: U.S. Geological Survey, 2008).
33. Saliba and Jacobs, "Saving the San Pedro River."

Chapter 9. That Sinking Feeling

Epigraph: G. Hardin, "The Tragedy of the Commons," *Science* 162, no. 3859 (1968).

1. E. Poland, G. P. Knight, and J. Poland, "Joe Poland and the USGS," in *Proceedings of the Dr. Joseph F. Poland Symposium on Land Subsidence,* ed. J. W. Borchers (Belmont, CA: Star Publishing, 1995), 3–5.
2. Ibid.
3. N. Hundley, *The Great Thirst: Californians and Water: A History* (Berkeley: University of California Press, 2001).
4. F. S. Riley, "Mechanics of Aquifer Systems—The Scientific Legacy of Joseph F. Poland," in *Proceedings of the Dr. Joseph F. Poland Symposium on*

Land Subsidence, ed. J. W. Borchers (Belmont, CA: Star Publishing, 1995), 13–27.

5. Poland, Knight, and Poland, "Joe Poland and the USGS."
6. D. Galloway and F. S. Riley, "San Joaquin Valley, California," in *Land Subsidence in the United States,* ed. D. Galloway, D. R. Jones, and S. E. Ingebritsen (Reston, VA: U.S. Geological Survey, 1999), 23–34.
7. D. Griffin and K. J. Anchukaitis, "How Unusual Is the 2012–2014 California Drought?" *Geophysical Research Letters* 41, no. 24 (2014).
8. M. Sneed, J. Brandt, and M. Solt, *Land Subsidence along the Delta-Mendota Canal in the Northern Part of the San Joaquin Valley, California, 2003–10* (Reston, VA: U.S. Geological Survey, 2013).
9. Galloway and Riley, "San Joaquin Valley, California."
10. Luhdorff & Scalmanini Consulting Engineers (LSCE) et al., *Land Subsidence from Groundwater Use in California* (Woodland: Prepared by LSCE with support by the California Water Foundation, 2014).
11. M. Sneed, "Recently Measured Rapid Land Subsidence in the San Joaquin Valley, California," presentation at 2015 NGWA Groundwater Summit, San Antonio, Texas, Mar. 18, 2015.
12. Sneed, Brandt, and Solt, *Land Subsidence along the Delta-Mendota Canal.*
13. C. C. Faunt and M. Sneed, "Water Availability and Subsidence in California's Central Valley," *San Francisco Estuary and Watershed Science* 13, no. 3 (2015); Michelle Sneed quoted in "Why is California Sinking? (Hint: Drought)?," *PBS Newshour,* Oct. 8, 2015, available at http://www.pbs.org/newshour/extra/daily_videos/why-is-california-sinking-hint-drought/ (accessed May 6, 2016).
14. LSCE et al., *Land Subsidence from Groundwater Use in California.*
15. C. Hauge et al., "Land Subsidence: Déjà Vu All Over Again," *Hydrovisions* 23, no. 4 (2014).
16. Quoted in I. James, V. Gibbons, and D. L. Taylor, "Calls Grow for More Oversight of California's Groundwater," *Desert Sun,* Dec. 21, 2013.
17. California History Center, *Water in the Santa Clara Valley: A History* (Cupertino, CA: De Anza College, 1981).
18. "Santa Clara Valley Water District," available at http://www.valleywater.org (accessed May 5, 2016); S. E. Ingebritsen and D. R. Jones, "Santa Clara Valley, California," in *Land Subsidence in the United States,* ed. D. Galloway, D. R. Jones, and S. E. Ingebritsen (Reston, VA: U.S. Geological Survey, 1999), 15–22.
19. Ingebritsen and Jones, "Santa Clara Valley."
20. Ibid.; C. F. Tolman and J. F. Poland, "Ground-Water, Salt-Water Infiltration, and Ground-Surface Recession in Santa Clara Valley, Santa Clara County, California," *Transactions, American Geophysical Union* 21, no. 1 (1940).

21. "Santa Clara Valley Water District."
22. J. Kane, "Galveston: The Mother of all U.S. Natural Disasters," *PBS Newshour,* Sept. 28, 2011; "The 1900 Galveston Hurricane," available at http://www.islandnet.com/~see/weather/events/1900hurr.htm (accessed May 6, 2016).
23. L. S. Coplin and D. Galloway, "Houston-Galveston, Texas," in *Land Subsidence in the United States,* ed. D. Galloway, D. R. Jones, and S. E. Ingebritsen (Reston, VA: U.S. Geological Survey, 1999), 35–48.
24. Ibid.; S. V. Stork and M. Sneed, *Houston-Galveston Bay Area, Texas, from Space—A New Tool for Mapping Land Subsidence* (Houston: U.S. Geological Survey, 2002).
25. Coplin and Galloway, "Houston-Galveston, Texas."
26. E. Cabral-Cano et al., "Space Geodetic Imaging of Rapid Ground Subsidence in Mexico City," *Geological Society of America Bulletin* 120, nos. 11–12 (2008).
27. S. Dillon, "Mexico City Journal; Capital's Downfall Caused by Drinking . . . of Water," *New York Times,* Jan. 29, 1998.
28. Quoted in ibid.
29. A. Hernández-Espriú et al., "The DRASTIC-Sg Model: An Extension to the DRASTIC approach for Mapping Groundwater Vulnerability in Aquifers Subject to Differential Land Subsidence, with Application to Mexico City," *Hydrogeology Journal* 22, no. 6 (2014).
30. IGRES Freshwater Resources Management Project, *Sustainable Groundwater Management in Asian Cities* (Hayama, Japan: Institute for Global Environmental Strategies [IGRES], 2007); S. Buapeng and S. Foster, *Controlling Groundwater Abstraction and Related Environmental Degradation in Metropolitan Bangkok—Thailand* (New York: The World Bank, 2008), available at http://siteresources.worldbank.org/INTWAT/Resources/GWMATE_CP_20_Bangkok.pdf (accessed May 5, 2016); T. Yamanaka, "Tracing a Confined Groundwater Flow System under the Pressure of Excessive Groundwater Use in the Lower Central Plain, Thailand," *Hydrological Processes* 25, no. 17 (2011).
31. L. E. Erban, S. M. Gorelick, and H. A. Zebker, "Groundwater Extraction, Land Subsidence, and Sea-Level Rise in the Mekong Delta, Vietnam," *Environmental Research Letters* 9 (2014).

Chapter 10. Recharge and Recycling

Epigraph: Van Vuuren is quoted in P. L. Du Pisani, "Surviving in an Arid Land: Direct Reclamation of Potable Water at Windhoek's Goreangab Reclamation Plant," *Arid Lands Newsletter* 56 (Nov.–Dec. 2004).

1. National Research Council, *Prospects for Managed Underground Storage of Recoverable Water* (Washington, DC: National Academies Press, 2008), 34.
2. M. T. Moreo and A. Swancar, *Evaporation from Lake Mead, Nevada and Arizona, March 2010 through February 2012* (Denver, CO: U.S. Geological Survey, 2013).
3. "Water Replenishment District of Southern California," available at http://www.wrd.org/ (accessed May 6, 2016).
4. B. Bourquard and C. Landry, "California Potable Reuse Gets a Boost with New Legislation," *Water Resources IMPACT* 16, no. 1 (2014); J. Ross, "A Win-Win Situation," *Groundwater Monitoring & Remediation* 34, no. 3 (2014).
5. "Orange County Water District," available at http://www.ocwd.com/ (accessed May 6, 2016).
6. Ibid.
7. F. Bloetscher et al., "Lessons Learned from Aquifer Storage and Recovery (ASR) Systems in the United States," *Journal of Water Resource and Protection* 6, no. 17 (2014).
8. National Research Council, *Water Reuse: Potential for Expanding the Nation's Water Supply through Reuse of Municipal Wastewater* (Washington, DC: National Academies Press, 2012); A. D. Levine and T. Asano, "Recovering Sustainable Water from Wastewater," *Environmental Science & Technology* 38, no. 11 (2004).
9. National Research Council, *Water Reuse;* F. Barringer, "As 'Yuck Factor' Subsides, Treated Wastewater Flows from Taps," *New York Times,* Feb. 9, 2012; B. Jiménez and T. Asano, "Water Reclamation and Reuse around the World," in *Water Reuse: An International Survey of Current Practice, Issues and Needs,* ed. B. Jiménez and T. Asano (London: IWA Publishing, 2008), 3–26.
10. L. Martin, "Looking to Singapore for Water Scarcity Solutions," *Water Online,* May 1, 2014.
11. National Research Council, *Water Reuse;* A. G. Harris and E. Tuttle, *Geology of National Parks* (Dubuque, Iowa: Kendall/Hunt, 1983), 6; D. G. Metzger, *Geology in Relation to Availability of Water along the South Rim Grand Canyon National Park Arizona* (Washington, DC: U.S. Geological Survey, 1961).
12. National Research Council, *Water Reuse.*
13. Ibid.
14. "About the Tampa Bay Estuary Program," available at http://www.tbep.org (accessed May 13, 2016).
15. A. R. Parker and C. R. Lawrence, "Water Capture by a Desert Beetle," *Nature* 414, no. 6859 (2001); "Namib Desert Beetle Inspires Self-Filling Water Bottle," *BBC News,* Nov. 23, 2012.

16. National Research Council, *Water Reuse;* I. B. Law, "Advanced Reuse—From Windhoek to Singapore and Beyond," *Water* 30, no. 5 (2003).
17. D. Weissmann, "Texas Town Closes the Toilet-to-Tap Loop: Is This Our Future Water Supply?," *Marketplace,* Jan. 6, 2014; P. P. Livingston and R. R. Bennett, *Geology and Ground-Water Resources of the Big Spring Area, Texas* (Washington, DC: U.S. Geological Survey, 1944).
18. Weissmann, "Texas Town"; L. Martin, "Texas Leads the Way with First Direct Potable Reuse Facilities in U.S.," *Water Online,* Sept. 16, 2014.
19. Martin, "Texas Leads the Way"; Associated Press, "Texas OKs Wastewater Reuse for Parched City," *Star-Telegram,* June 27, 2014; B. Hanna, "Wichita Falls Says Goodbye to Potty Water for Now," *Star-Telegram,* July 27, 2015.
20. D. W. Kolpin et al., "Pharmaceuticals, Hormones, and Other Wastewater Contaminants in U.S. Streams, 1999–2000: A National Reconnaissance," *Environmental Science & Technology* 36, no. 6 (2002).
21. T. Heberer et al., "Occurrence and Fate of Pharmaceuticals during Bank Filtration—Preliminary Results from Investigations in Germany and the United States," *Journal of Contemporary Water Research & Education* 120, no. 1 (2000); B. Kasprzyk-Hordern et al., "The Removal of Pharmaceuticals, Personal Care Products, Endocrine Disruptors and Illicit Drugs during Wastewater Treatment and Its Impact on the Quality of Receiving Waters," *Water Research* 43, no. 2 (2009).
22. T. Heberer et al., "Field Studies on the Fate and Transport of Pharmaceutical Residues in Bank Filtration," *Ground Water Monitoring & Remediation* 24, no. 2 (2004).
23. K. M. Hiscock and T. Grischek, "Attenuation of Groundwater Pollution by Bank Filtration," *Journal of Hydrology* 266, nos. 3–4 (2002).
24. Heberer et al., "Occurrence and Fate of Pharmaceuticals."
25. D. A. Burgard et al., "Working Upstream: How Far Can You Go with Sewage-Based Drug Epidemiology?," *Environmental Science & Technology* 48, no. 3 (2014).

Chapter 11. Poisoning the Well

Epigraph: Franklin, "Protection of Towns from Fire," *Pennsylvania Gazette,* Feb. 4, 1735.

1. Council on Environmental Quality (CEQ), *Contamination of Ground Water by Toxic Organic Chemicals* (Washington, DC: CEQ, 1981); P. W. Hadley and C. J. Newell, "Groundwater Remediation: The Next 30 Years," *Ground Water* 50, no. 5 (2012).
2. D. M. Mackay and J. A. Cherry, "Groundwater Contamination: Pump-and-Treat Remediation," *Environmental Science & Technology* 23, no. 6 (1989).

3. "Technical Tour Book for Lake Erie to Lake Ontario, Spills, Mills, and Landfills, and GW/SW Glacial Geology," International Association of Hydrogeologists 2012 Congress, Niagara Falls, Sept. 19, 2012; R. G. McLaren et al., "Numerical Simulation of DNAPL Emissions and Remediation in a Fractured Dolomitic Aquifer," *Journal of Contaminant Hydrology* 136–137 (2012).
4. J. Steinberg, "Water Regulators Approve Comprehensive Clean-up for Hinkley Plume," *San Bernardino Sun,* Nov. 5, 2015.
5. J. Steinberg, "Hinkley: No Hollywood Ending for Erin Brockovich's Tainted Town," *San Bernardino Sun,* July 8, 2013.
6. I. James, "Calif. Asked to Weigh What Chromium-6 Standard is Worth," *Desert Sun,* Oct. 5, 2013.
7. Ibid.; J. A. Izbicki et al., "Cr(VI) Occurrence and Geochemistry in Water from Public-Supply Wells in California," *Applied Geochemistry* 63 (2015).
8. J. F. Pankow and J. A. Cherry, *Dense Chlorinated Solvents and Other DNAPLs in Groundwater* (Portland, OR: Waterloo Press, 1996).
9. National Research Council, *Alternatives for Managing the Nation's Complex Contaminated Groundwater Sites* (Washington, DC: National Academies Press, 2012), 11.
10. F. Schwille, *Dense Chlorinated Solvents in Porous and Fractured Media: Model Experiments,* trans. from the German by J. F. Pankow (Boca Raton, FL: Lewis, 1988).
11. Pankow and Cherry, *Dense Chlorinated Solvents;* R. E. Jackson, "Recognizing Emerging Environmental Problems: The Case of Chlorinated Solvents in Groundwater," *Technology and Culture* 45, no. 1 (2004).
12. Pankow and Cherry, *Dense Chlorinated Solvents,* 19.
13. National Research Council, *Groundwater Contamination;* Council on Environmental Quality, *Contamination of Ground Water.*
14. R. E. Doherty, "A History of the Production and Use of Carbon Tetrachloride, Tetrachloroethylene, Trichloroethylene and 1,1,1-Trichloroethane in the United States. Part 1: Historical Background; Carbon Tetrachloride and Tetrachloroethylene," *Journal of Environmental Forensics* 1, no. 2 (2000).
15. A. O'Hanlon, "Dry-Cleaning: 'PERC' + Water Does Wonders," *Washington Post,* Nov. 12, 1997.
16. State Coalition for the Remediation of Dry Cleaners, *Newsletter* (Dec. 2010).
17. National Research Council, *Alternatives.*
18. J. Gelt et al., *Water in the Tucson Area: Seeking Sustainability* (Tucson: Arizona Water Resources Research Center, 1999), 65–67; M. L. Brusseau and M. Narter, "Assessing the Impact of Chlorinated-Solvent Sites on Metropolitan Groundwater Resources," *Groundwater* 51, no. 6 (2013).

19. T. K. G. Mohr, *Environmental Investigation and Remediation: 1,4-Dioxane and other Solvent Stabilizers* (Boca Raton, FL: CRC Press, 2010).
20. Interstate Technology & Regulatory Council (ITRC), *Vapor Intrusion Pathway: A Practical Guideline. VI-1* (Washington, DC: ITRC, 2007), available at http://www.itrcweb.org/Guidance/ListDocuments?topicID=28&subTopicID=39 (accessed May 8, 2016).
21. M. L. Wald, "Agency Will Ask Congress to Drop Gasoline Additive," *New York Times,* July 27, 1999.
22. Ibid.; M. J. Moran, J. S. Zogorski, and P. J. Squillace, *Occurrence and Implications of Methyl tert-Butyl Ether and Gasoline Hydrocarbons in Ground Water and Source Water in the United States and in Drinking Water in Twelve Northeast and Mid-Atlantic States, 1993–2002* (Reston, VA: U.S. Geological Survey, 2004).
23. J. M. McDade et al., "Exceptionally Long MTBE Plumes of the Past Have Greatly Diminished," *Groundwater* 53, no. 4 (2015).
24. U.S. Environmental Protection Agency, *Drinking Water Advisory—Consumer Acceptability Advice and Health Effects Analysis on Methyl Tertiary-butyl Ether* (Washington, DC: U.S. Environmental Protection Agency, 1997).
25. J. S. Zogorski et al., *The Quality of Our Nation's Waters—Volatile Organic Compounds in the Nation's Ground Water and Drinking Water Supply Wells* (Reston, VA: U.S. Geological Survey, 2006).
26. U.S. Environmental Protection Agency, *Achieving Clean Air and Clean Water: The Report of the Blue Ribbon Panel on Oxygenates in Gasoline* (Washington, DC: U.S. Environmental Protection Agency, 1999); U.S. Environmental Protection Agency, *State Actions Banning MTBE (Statewide)* (Washington, DC: U.S. Environmental Protection Agency, 2004).
27. Mackay and Cherry, "Groundwater Contamination."
28. B. L. Parker, J. A. Cherry, and S. W. Chapman, "Discrete Fracture Network Approach for Studying Contamination in Fractured Rock," *AQUA Mundi: Journal of Water Science* 60 (2012).
29. S. S. D. Foster, "The Chalk Groundwater Tritium Anomaly—A Possible Explanation," *Journal of Hydrology* 25, nos. 1–2 (1975).
30. National Research Council, *Alternatives.*
31. M. C. Kavanaugh et al., *The DNAPL Remediation Challenge: Is There a Case for Source Depletion?* (Washington, DC: U.S. Environmental Protection, 2003).
32. D. Chakraborti, B. Das, and M. T. Murrill, "Examining India's Groundwater Quality Management," *Environmental Science & Technology* 45, no. 11

(2011); J. Kaiman, "China Says More Than Half of Its Groundwater Is Polluted," *Guardian,* Apr. 23, 2014.

33. John Cherry, personal communication.

Chapter 12. Pathogens

Epigraph: Ki-moon is quoted in "Dirty Water Kills More People than War, UN Says," *Telegraph*, Mar. 23, 2010.

1. S. Hempel, "John Snow," *Lancet* 381, no. 9874 (2013).
2. G. F. Craun et al., "Causes of Outbreaks Associated with Drinking Water in the United States from 1971 to 2006," *Clinical Microbiology Reviews* 23, no. 3 (2010); M. A. Borchardt et al., "Incidence of Enteric Viruses in Groundwater from Household Wells in Wisconsin," *Applied Environmental Microbiology* 69, no. 2 (2003); National Ground Water Association recommendations available at WellOwner.org (accessed May 8, 2016).
3. Expert Panel on Groundwater, *The Sustainable Management of Groundwater in Canada* (Ottawa: The Council of Canadian Academies, 2009); J. Brooke, "Few Left Untouched after Deadly *E.coli* Flows through an Ontario Town's Water," *New York Times,* July 10, 2000.
4. Expert Panel on Groundwater, *Sustainable Management.*
5. Brooke, "Few Left Untouched."
6. Expert Panel on Groundwater, *Sustainable Management,* 12.
7. M. A. Borchardt et al., "Viruses in Non-Disinfected Drinking Water from Municipal Wells and Community Incidence of Acute Gastrointestinal Illness," *Environmental Health Perspectives* 120, no. 9 (2012).
8. K. R. Bradbury et al., "Source and Transport of Human Enteric Viruses in Deep Municipal Water Supply Wells," *Environmental Science & Technology* 47, no. 9 (2013); M. A. Borchardt et al., "Human Enteric Viruses in Groundwater from a Confined Bedrock Aquifer," *Environmental Science & Technology* 41, no. 18 (2007).
9. Bradbury et al., "Source and Transport of Human Enteric Viruses."
10. Ibid.
11. U.S. Environmental Protection Agency, "National Primary Drinking Water Regulations: Ground Water Rule, Final Rule," *Federal Register* 71, no. 216 (2006).
12. Borchardt et al., "Viruses in Non-Disinfected Drinking Water."
13. Ibid.: R. J. Hunt et al., "Assessment of Sewer Source Contamination of Drinking Water Wells Using Tracers and Human Enteric Viruses," *Environmental Science & Technology* 44, no. 20 (2010).
14. U.S. Environmental Protection Agency, "National Primary Drinking Water Regulations."

15. S. Elbow, "GOP Proposes Rollback of Mandatory Disinfection for Drinking Water," *Cap Times,* Feb. 23, 2011.
16. Ibid.; B. Hulsey et al., "Letter from Eight Wisconsin State Legislatures to Lisa Jackson, Administrator, Environmental Protection Agency," Wisconsin State Legislature, June 1, 2012; M. A. Borchardt, "Testimony to Wisconsin Senate Regarding SB19 and AB23," Mar. 29, 2011, both in authors' possession.

Chapter 13. Arsenic

Epigraph: Ozonoff is quoted in D. Fagin, *Toms River: A Story of Science and Salvation* (New York: Bantam Books, 2013), 442.

1. R. Smith, "Arsenic: A Murderous History," available at http://www.dartmouth.edu/~toxmetal/arsenic/history.html (accessed May 8, 2016).
2. Ibid.
3. P. L. Smedley and D. G. Kinniburgh, "A Review of the Source, Behaviour and Distribution of Arsenic in Natural Waters," *Applied Geochemistry* 17, no. 2 (2002).
4. A. H. Smith et al., "Arsenic Epidemiology and Drinking Water Standards," *Science* 296, no. 5576 (2002); C. W. Schmidt, "Low-Dose Arsenic: In Search of a Risk Threshold," *Environmental Health Perspectives* 122, no. 5 (2014).
5. Smedley and Kinniburgh, "A Review of the Source, Behaviour and Distribution of Arsenic in Natural Waters"; A. H. Smith, E. O. Lingas, and M. Rahman, "Contamination of Drinking-Water by Arsenic in Bangladesh: A Public Health Emergency," *Bulletin World Health Organization* 78, no. 9 (2000).
6. C. H. Atkins, "Development of the National Water Supply and Sanitation Program in India," *American Journal of Public Health and the Nation's Health* 47, no. 10 (1957).
7. D. Chakraborti, B. Das, and M. T. Murrill, "Examining India's Groundwater Quality Management," *Environmental Science and Technology* 45, no. 1 (2011); C. F. Harvey et al., "Groundwater Dynamics and Arsenic Contamination in Bangladesh," *Chemical Geology* 228, nos. 1–3 (2006).
8. Smith, Lingas, and Rahman, "Contamination of Drinking Water."
9. Ibid.; Chakraborti, Das, and Murrill, "Examining India's Groundwater Quality Management."
10. W. G. Burgess et al., "Vulnerability of Deep Groundwater in the Bengal Aquifer System to Contamination by Arsenic," *Nature Geoscience* 3, no. 2 (2010).
11. Smith, Lingas, and Rahman, "Contamination of Drinking Water."
12. R. B. Johnston, S. Hanchett, and M. H. Khan, "The Socio-Economics of Arsenic Removal," *Nature Geoscience* 3, no. 1 (2010).

13. Burgess et al., "Vulnerability of Deep Groundwater"; H. A. Michael and C. I. Voss, "Evaluation of the Sustainability of Deep Groundwater as an Arsenic-Safe Resource in the Bengal Basin," *Proceedings, National Academy of Sciences USA* 105, no. 25 (2008).
14. S. Fendorf, H. A. Michael, and A. van Geen, "Spatial and Temporal Variations of Groundwater Arsenic in South and Southeast Asia," *Science* 328, no. 5982 (2010).
15. Harvey et al., "Groundwater Dynamics and Arsenic Contamination in Bangladesh."
16. R. B. Neumann et al., "Anthropogenic Influences on Groundwater Arsenic Concentrations in Bangladesh," *Nature Geoscience* 3, no. 1 (2010); J. W. Stuckey et al., "Arsenic Release Metabolically Limited to Permanently Water-Saturated Soil in Mekong Delta," *Nature Geoscience* (Nov. 30, 2015).
17. P. Ravenscroft, H. Brammer, and K. Richards, *Arsenic Pollution: A Global Synthesis* (Chichester, Eng.: Wiley-Blackwell, 2009).
18. L. E. Erban et al., "Release of Arsenic to Deep Groundwater in the Mekong Delta, Vietnam, Linked to Pumping-Induced Land Subsidence," *Proceedings, National Academy of Sciences USA* 110, no. 34 (2013).
19. National Research Council, *Arsenic in Drinking Water* (Washington, DC: National Academies Press, 1999).
20. National Research Council, *Arsenic in Drinking Water: 2001 Update* (Washington, DC: National Academies Press, 2001); Albuquerque Bernalillo County Water Utility Authority, "Water Resources Education," available at http://www.abcwua.org/education/31_Arsenic.html (accessed May 8, 2016); O. Reed, "Pilot Program Could Bring Some Defunct City Wells Back Online," *Albuquerque Journal,* Sept. 15, 2015.
21. Schmidt, "Low-Dose Arsenic."
22. National Research Council, *Critical Aspects of EPA's IRIS Assessment of Inorganic Arsenic: Interim Report* (Washington, DC: National Academies Press, 2013); M. F. Hughes et al., "Arsenic Exposure and Toxicology: A Historical Perspective," *Toxicological Sciences* 123, no. 2 (2011).
23. M. A. Maupin et al., *Estimated Use of Water in the United States in 2010* (Reston, VA: U.S. Geological Survey, 2014).
24. M. Borsuk et al., *Arsenic in Private Wells in NH, Year 2 Final Report, Public Health Contract, Annual Performance Report, CDC Grant #1U53/EH001110–01* (Hanover, NH: Dartmouth Toxic Metals Superfund Research Program, 2015); J. D. Ayotte et al. "Arsenic in Groundwater in Eastern New England: Occurrence, Controls, and Human Health Implications," *Environmental Science & Technology* 37, no. 10 (2003).

25. Borsuk et al., *Arsenic in Private Wells.*
26. G. A. Wasserman et al., "A Cross-Sectional Study of Well Water Arsenic and Child IQ in Maine Schoolchildren," *Environmental Health* 13, no. 1 (2014).
27. Y. Zheng and J. D. Ayotte, "At the Crossroads: Hazard Assessment and Reduction of Health Risks from Arsenic in Private Well Waters of the Northeastern United States and Atlantic Canada," *Science of the Total Environment* 505 (2015).
28. Agency for Toxic Substances and Disease Registry, "Priority List of Hazardous Substances," available at http://www.atsdr.cdc.gov/spl (accessed May 5, 2016).
29. American Academy of Pediatrics, Committee on Environmental Health and Committee on Infectious Diseases, "Drinking Water from Private Wells and Risks to Children," *Pediatrics* 123, no. 6 (2009).
30. J. Fawell et al., ed., *Fluoride in Drinking Water* (London: World Health Organization, IWA Publishing, 2006).
31. Ibid.; S. Ayoob, and A. K. Gupta, "Fluoride in Drinking Water: A Review on the Status and Stress Effects," *Critical Reviews in Environmental Science and Technology* 36, no. 6 (2006).
32. L. A. DeSimone, P. B. McMahon, and M. R. Rosen, *Water Quality in Principal Aquifers of the United States, 1991–2010* (Reston, VA: U.S. Geological Survey, 2014).
33. Z. Szabo et al., *Relation of Distribution of Radium, Nitrate, and Pesticides to Agricultural Land Use and Depth, Kirkwood-Cohansey Aquifer System, New Jersey Coastal Plain, 1990–1991* (West Trenton: U.S. Geological Survey, 1997).
34. B. C. Jurgens et al., "Effects of Groundwater Development on Uranium—Central Valley, California, USA," *Ground Water* 48, no. 6 (2010).

Chapter 14. Fracking

Epigraph: Getty is quoted in BrainyQuote.com, Xplore Inc, http://www.brainyquote.com/quotes/quotes/j/jpaulgett100065.html (accessed May 13, 2016).

1. R. Showstack, "Senate Forum on Shale Gas Development Explores Environmental and Industry Issues," *Eos* 94, no. 23 (2013).
2. R. Gold, *The Boom: How Fracking Ignited the American Energy Revolution and Changed the World* (New York: Simon & Schuster, 2014).
3. R. D. Vidic et al., "Impact of Shale Gas Development on Regional Water Quality," *Science* 340, no. 6134 (2013).
4. Energy Information Administration, *United States Annual Energy Outlook, 2015* (Washington, DC: U.S. Department of Energy, 2015).

5. M. Dusseault and R. Jackson, "Seepage Pathway Assessment for Natural Gas to Shallow Groundwater during Well Stimulation, in Production, and after Abandonment," *Environmental Geosciences* 21, no. 3 (2014).
6. E. Hand, "Injection Wells Blamed in Oklahoma Earthquakes," *Science* 345, no. 6192 (2014).
7. B. D. Drollette et al., "Elevated Levels of Diesel Range Organic Compounds in Groundwater near Marcellus Gas Operations Are Derived from Surface Activities," *Proceedings, National Academy of Sciences USA* 112, no. 43 (2015).
8. Dusseault and Jackson, "Seepage Pathway Assessment for Natural Gas."
9. R. J. Davies et al., "Oil and Gas Wells and Their Integrity: Implications for Shale and Unconventional Resource Exploitation," *Marine and Petroleum Geology* 56 (2014).
10. S. Bachu and R. L. Valencia, "Well Integrity: Challenges and Risk Mitigation Measures," *Bridge* (Summer 2014); R. E. Jackson et al., "Groundwater Protection and Unconventional Gas Extraction: The Critical Need for Field-Based Hydrogeological Research," *Groundwater* 51 no. 4 (2013).
11. Dusseault and Jackson, "Seepage Pathway Assessment"; K. Muehlenbachs, *Identifying the Sources of Fugitive Methane Associated with Shale Gas Development* (Washington, DC: Resources for the Future, 2011).
12. Vidic et al., "Impact of Shale Gas Development on Regional Water Quality."
13. Ibid.
14. S. G. Osborn et al., "Methane Contamination of Drinking Water Accompanying Gas-Well Drilling and Hydraulic Fracturing," *Proceedings, National Academy of Sciences USA* 108, no. 20 (2011); L. J. Molofsky et al., "Evaluation of Methane Sources in Groundwater in Northeastern Pennsylvania," *Ground Water* 51, no. 3 (2013); D. I. Siegel et al., "Methane Concentrations in Water Wells Unrelated to Proximity to Existing Oil and Gas Wells in Northeastern Pennsylvania," *Environmental Science & Technology* 49, no. 9 (2015); T. H. Darrah et al., "Noble Gases Identify the Mechanisms of Fugitive Gas Contamination in Drinking-Water Wells Overlying the Marcellus and Barnett Shales," *Proceedings, National Academy of Sciences USA* 111, no. 39 (2014).
15. R. E. Jackson et al., "Groundwater Protection and Unconventional Gas Extraction: The Critical Need for Field-Based Hydrogeological Research," *Groundwater* 51 no. 4 (2013); Expert Panel on Harnessing Science and Technology to Understand the Environmental Impacts of Shale Gas Extraction, *Environmental Impacts of Shale Gas Extraction in Canada* (Ottawa: Council of Canadian Academies, 2014); D. J. Soeder, "Adventures

in Groundwater Monitoring: Why Has It Been So Difficult to Obtain Groundwater Data Near Shale Gas Wells?" *Environmental Geosciences* 22, no. 4 (2015).

Chapter 15. Nitrate and Aquifer Protection

Epigraph: Stern is quoted in K. Weekes, *Women Who Knew Everything!: 3,241 Quips, Quotes, and Brilliant Remarks* (Philadelphia: Quirk Books, 2007), 305.

1. S. S. Seacrest, "The History of the Groundwater Foundation," *Aquifer* 15, no. 2 (2000).
2. C. Kreifels, "The Little Idea That Grew: Celebrating 20 Years of Educational Experience," *Aquifer* 20, no. 2 (2005).
3. Ibid.; "Mother Turned Clean Groundwater Advocate," *Aquifer* 22, no. 2 (2007).
4. J. Wendel, "Gulf of Mexico Dead Zone Largest since 2002," *Eos*, Sept. 15, 2015.
5. R. F. Service, "New Recipe Produces Ammonia from Air, Water, and Sunlight," *Science* 345, no. 6197 (2014).
6. "The 50 Greatest Breakthroughs since the Wheel," *Atlantic Magazine* (Nov. 2013); V. Smil, *Enriching the Earth: Fritz Haber, Carl Bosch, and the Transformation of World Food Production* (Cambridge, MA: MIT Press, 2004); N. M. Dubrovsky et al., *Nutrients in the Nation's Streams and Groundwater, 1992–2004* (Reston, VA: U.S. Geological Survey, 2003).
7. G. E. Fogg and E. M. LaBolle, "Motivation of Synthesis, with an Example on Groundwater Quality Sustainability," *Water Resources Research* 42, no. 3 (2006).
8. T. Harter et al., *Addressing Nitrate in California's Drinking Water* (Davis: Center for Watershed Sciences, University of California, 2012), 10.
9. "NAE Grand Challenges for Engineering," available at http://www.engineeringchallenges.org/cms/8996/9132.aspx (accessed May 5, 2016).
10. S. W. Phillips and B. D. Lindsey, *The Influence of Ground Water on Nitrogen Delivery to the Chesapeake Bay* (Baltimore: U.S. Geological Survey, 2003).
11. B. D. Lindsey et al., *Residence Times and Nitrate Transport of Ground Water Discharging to Streams in the Chesapeake Bay Watershed* (Denver: U.S. Geological Survey, 2003); W. E. Sanford and J. P. Pope, "Quantifying Groundwater's Role in Delaying Improvements to Chesapeake Bay Water Quality," *Environmental Science & Technology* 47, no. 23 (2013).
12. Harter et al., *Addressing Nitrate in California's Drinking Water.*
13. Ibid.; J. Miller, "California Farm Communities Suffer Tainted Drinking Water," *High Country News*, June 24, 2013.
14. Ibid.

15. Quoted in N. Kresic, *Groundwater Resources: Sustainability, Management, and Restoration* (New York: McGraw-Hill, 2009), 619.
16. "Source Water Collaborative," available at http://www.sourcewater collaborative.org/ (accessed May 5, 2016).
17. L. F. Jørgensen and J. Stockmarr, "Groundwater Monitoring in Denmark: Characteristics, Perspectives, and Comparison with Other Countries," *Hydrogeology Journal* 17, no. 4 (2009); R. Thomsen, V. H. Søndergaard, and K. I. Sørensen, "Hydrogeological Mapping as a Basis for Establishing Site-Specific Groundwater Protection Zones in Denmark," *Hydrogeology Journal* 12, no. 5 (2004).
18. J. C. Refsgaard et al., "Groundwater Modeling in Integrated Water Resources Management—Visions for 2020," *Ground Water* 48, no. 5 (2010).
19. R. Thomsen, V. H. Søndergaard, and P. Klee, "Greater Water Security with Groundwater—Groundwater Mapping, and Sustainable Groundwater Management," available at www.rethinkwater.dk (accessed May 5, 2016).
20. National Research Council, *Ground Water Vulnerability Assessment* (Washington, DC: National Academy Press, 1993), 2. The council called this the "First Law of Groundwater Vulnerability."
21. K. J. Pieper et al., "Profiling Private Water Systems to Identify Patterns of Waterborne Lead Exposure," *Environmental Science & Technology* 49, no. 21 (2015).

Chapter 16. Transboundary Aquifers

Epigraph: Brown is quoted in N. Hundley, *The Great Thirst: Californians and Water: A History* (Berkeley: University of California Press, 2001), 289.

1. J. M. Safford, "The Water Supply of Memphis," *State Board of Health Bulletin* 5, no. 7 (1890).
2. T. Charlier, "Memphis Taps into Desoto County's Well Levels," *Commercial Appeal,* Nov. 16, 1998; M. E. Campana, "Growing Groundwater Governance in Graceland: The Memphis Sand Aquifer," presentation at AWRA Annual Conference, Portland, Oregon, Nov. 4–7, 2013; M. A. Maupin et al., *Estimated Use of Water in the United States in 2010* (Reston, VA: U.S. Geological Survey, 2014).
3. B. Waldron and D. Larsen, "Pre-Development Groundwater Conditions Surrounding Memphis, Tennessee: Controversy and Unexpected Outcomes," *Journal of the American Water Resources Association* 51, no. 1 (2015); "Is Memphis Stealing Water from Mississippi?" *Jackson Clarion-Ledger,* Oct. 6, 2014.
4. S. Puri and A. Aureli, eds., *Atlas of Transboundary Aquifers* (Paris, UNESCO, 2009).

5. G. de los Cobos, "The Transboundary Aquifer of the Geneva Region [Switzerland and France]: A Successful 30-Year Management between the State of Geneva and French Border Communities," in *International Conference "Transboundary Aquifers: Challenges and New Directions" (ISARM2010)* (Paris: UNESCO, 2010), 103.
6. G. Eckstein, "The Newest Transboundary Aquifer Agreement: Jordan and Saudi Arabia Cooperate over the Al-Sag/Al-Disi Aquifer," International Water Law Project Blog, available at www.internationalwaterlaw.org/blog (accessed May 5, 2016).
7. F. Sindico, "The Guarani Aquifer System and the Law on Transboundary Aquifers," *International Community Law Review* 13, no. 3 (2011).
8. B. J. Zebarth et al., "Groundwater Monitoring to Support Development of BMPs for Groundwater Protection: The Abbotsford-Sumas Aquifer Case Study," *Groundwater Monitoring & Remediation* 35, no. 1 (2015).
9. A. Wolf, "A Long Term View of Water and International Security," *Journal of Contemporary Water Research & Education* 142, no. 1 (2009).
10. M. Giordano et al., "A Review of the Evolution and State of Transboundary Freshwater Treaties," *International Environmental Agreements: Politics, Law and Economics* 14, no. 3 (2014).
11. Good Neighbor Environmental Board, "A Blueprint for Action on the U.S.-Mexico Border," Thirteenth Report of the Good Neighbor Environmental Board to the President and Congress of the United States, 2010 available at https://www.epa.gov/sites/production/files/documents/eng_gneb_13th_report_final.pdf (accessed May 5, 2016).
12. G. E. Eckstein, "Rethinking Transboundary Ground Water Resources Management: A Local Approach along the Mexico-U.S. Border," *Georgetown International Environmental Law Review* 25, no. 1 (2013).
13. S. B. Megdal and C. A. Scott, "The Importance of Institutional Asymmetries to the Development of Binational Aquifer Assessment Programs: The Arizona-Sonora Experience," *Water* 3, no. 3 (2011); Z. Sheng et al., "Mesilla Basin/Conejos-Médanos Section of the Transboundary Aquifer Assessment Program," in *Five-Year Interim Report of the United States–Mexico Transboundary Aquifer Assessment Program: 2007–2012*, ed. W. M. Alley (Reston, VA: U.S. Geological Survey, 2013).
14. S. Dibble, "Wetlands Become a Focus in Debate over Canal Lining," *U-T San Diego*, June 5, 2005.

Chapter 17. Sharing the Common Pool

Epigraph: Rowland is quoted in D. Wuebbles and J. Melillo, "Climate Change and Our Nation," *Eos*, June 1, 2015.

1. Quoted in C. McGlade, "The Battle for Water When the Wells Run Dry," *AZcentral.com*, June 7, 2015.
2. World Commission on Environment and Development, *Our Common Future* (New York: Oxford University Press, 1987).
3. S. Foster and H. Garduño, "Groundwater-Resource Governance: Are Governments and Stakeholders Responding to the Challenge?," *Hydrogeology Journal* 21, no. 2 (2013).
4. S. Jasechko and R. G. Taylor, "Intensive Rainfall Recharges Tropical Groundwaters," *Environmental Research Letters* 10, no. 12 (2015).
5. J. Diamond, *Collapse: How Societies Choose to Fail or Succeed* (New York: Penguin, 2005), 107–109.

Further Reading

There are many sources of information on groundwater.
We list some useful Internet sites here.

Basic Information on Groundwater

U.S. Geological Survey (http://water.usgs.gov/ogw/)
UK Groundwater Forum (http://www.groundwateruk.org/)
National Ground Water Association (www.ngwa.org)
Groundwater Foundation (http://www.groundwater.org/)
Ground Water Protection Council (http://www.gwpc.org/)

Sites with Daily News

WaterWired (http://aquadoc.typepad.com/waterwired/)
Circle of Blue (http://www.circleofblue.org/waternews/)
Aquafornia (http://www.watereducation.org/aquafornia)

Index